T0396741

LIGHT-INDUCED DEFECTS IN SEMICONDUCTORS

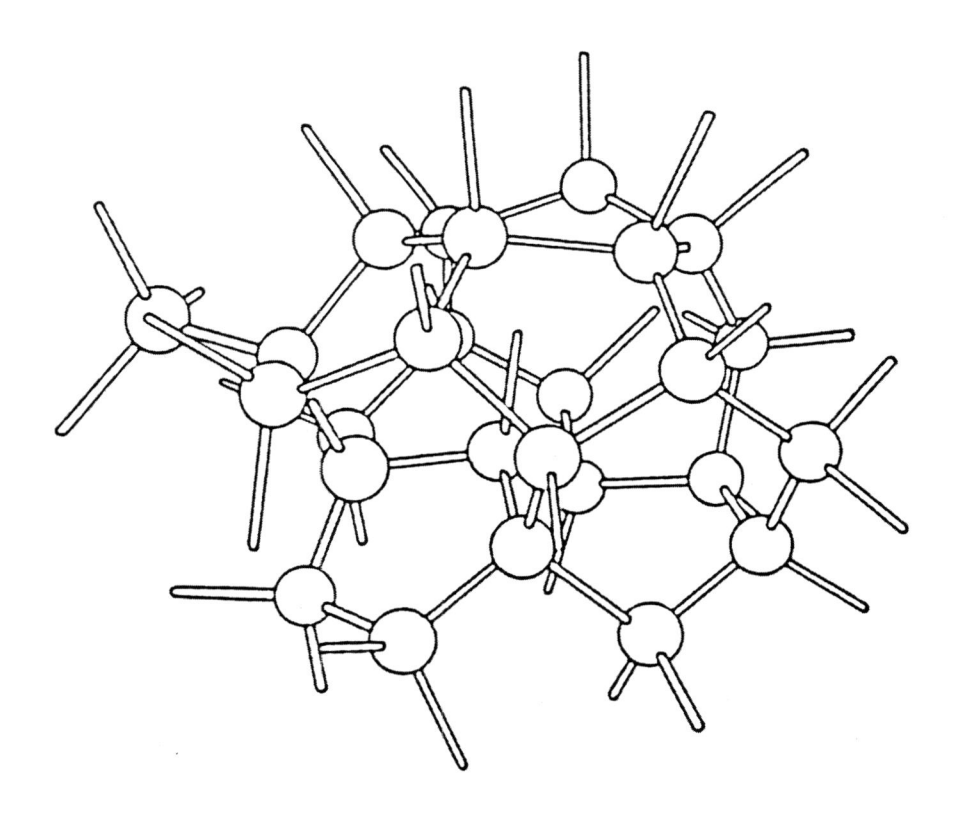

LIGHT-INDUCED DEFECTS IN SEMICONDUCTORS

Kazuo Morigaki
Harumi Hikita
Chisato Ogihara

PAN STANFORD PUBLISHING

Published by

Pan Stanford Publishing Pte. Ltd.
Penthouse Level, Suntec Tower 3
8 Temasek Boulevard
Singapore 038988

Email: editorial@panstanford.com
Web: www.panstanford.com

British Library Cataloguing-in-Publication Data
A catalogue record for this book is available from the British Library.

Light-Induced Defects in Semiconductors

ISBN 978-981-4411-48-6 (Hardcover)
ISBN 978-981-4411-49-3 (eBook)

Printed in the USA

Contents

Preface

The book deals with electronic and structural properties of light-induced defects, their light-induced creation processes, and related phenomena in crystalline, amorphous, and microcrystalline semiconductors. Recombination-enhanced defect reaction (REDR) has received much attention in connection with degradation of light-emitting diodes. Theoretical and experimental investigations relating to this issue have been extensively performed as discussed in Chapter 2, where we concentrate on REDR in GaAs and related materials. Light-induced defects in hydrogenated polycrystalline silicon are also treated in this chapter. The topics discussed in detail in Chapters 3 and 4 present our own investigations on hydrogenated amorphous silicon and hydrogenated micro-crystalline silicon, respectively. The results on light-induced defects obtained from elsewhere are also presented in these chapters. Models of light-induced defect creation in hydrogenated amorphous silicon are presented separatelyin Chapter 3 as this issue has been investigated by many authors in connection with light-induced degradation of amorphous silicon solar cells. Light-induced phenomena in amorphous chalcogenides have received much attentionboth from a fundamental point of view and for their applications. These phenomena and related models are summarized in Chapter 5. We hope that the book will be useful for students and researchers interested in all the above topics.

We thank S. Kugler, K. Murayama, C. Niikura, P. Roca i Cabarrocas, J. Singh, K. Shimakawa, I. Solomon, and K. Takeda, discussions with whom were very beneficial in the development of this book.

Kazuo Morigaki
Harumi Hikita
Chisato Ogihara
Summer 2014

Acknowledgment

We wish to acknowledge the permission of the following publishers for reproducing those figures (copyright materials) published in the journals and books whose sources are given in each figure: American Institute of Physics; American Physical Society; Cambridge University Press; Elsevier; European Photovoltaic Solar Energy Conference, WIP; John Wiley & Sons; Journal of Optoelectronic and Advanced Materials, INFIM; Kotai Butsuri, Agne Gijutsu Center; Materials Research Society; Plenum Press; Physical Society of Japan; Springer; Taylor & Francis; The Japan Society of Applied Physics.

Chapter 1

Introduction

1.1 Light-Induced Defect Creation

Light-induced defect creation in semiconductors is one of interesting and important issues involving recombination processes from the viewpoints of the physics of semiconductors as well as their applications such as solar cells and other opto-electronic devices. In this book, we first deal with recombination-enhanced defect reaction in crystalline semiconductors, particularly in GaAs and related materials. Recombination of excess carriers, either photogenerated or injected through a p-n junction, is of either radiative or nonradiative nature. Nonradiative recombination is involved in recombination-enhanced defect reaction, because emitted phonons associated with nonradiative recombination supply energy necessary for defect reaction. Thus, electron–phonon interaction plays an important role in recombination-enhanced defect reaction. The usual procedure to treat electron–phonon interaction uses a configuration coordinate model [Mott, 1978; Toyozawa, 2003]. The recombination-enhanced defect reaction is considered on the basis of the configuration coordinate model described in Section 2.2.3. Light-induced defects in crystalline

Light-Induced Defects in Semiconductors
Kazuo Morigaki, Harumi Hikita, and Chisato Ogihara
Copyright © 2015 Pan Stanford Publishing Pte. Ltd.
ISBN 978-981-4411-48-6 (Hardcover), 978-981-4411-49-3 (eBook)
www.panstanford.com

semiconductors have been investigated in III–V compounds, II–VI compounds, oxide materials such as silicon dioxides, and so on, but in this book, we concentrate on GaAs and related materials, as mentioned above. For other crystalline semiconductors, the reader is referred to [Redfield and Bube, 1996] for II–VI compounds, and to [Morigaki, 1999; Uchino *et al.*, 2001, 2003] for silicon dioxides. Light-induced defect creation in hydrogenated polycrystalline silicon is also considered for comparison with amorphous and microcrystalline hydrogenated silicon in Section 2.3.

For hydrogenated amorphous silicon (a-Si:H), hydrogenated microcrystalline silicon (μc-Si:H), and amorphous chalcogenides (a-Ch), we present their light-induced defect creation in Chapters 3, 4, and 5, respectively.

Light-induced defects and recombination-enhanced defect reaction in crystalline semiconductors are also reviewed by Lang [1992], Mooney [1992], and Redfield and Bube [1996]. For a-Si:H, the reader is referred to [Shimakawa *et al.*, 1995; Morigaki, 1999; Fritzsche, 2001; Singh and Shimakawa, 2003; Shimizu, 2004] and, for a-Ch, to [Shimakawa *et al.*, 1995; Morigaki, 1999; Singh and Shimakawa, 2003; Shimakawa *et al.*, 2004; Tanaka and Shimakawa, 2011].

1.2 Defects

Defects in crystals have been extensively investigated from the viewpoint of physics, historically for the first time on alkali halides since 1920s [Mott and Gurney, 1940]. Furthermore, it has been well known that Seitz [1954] reviewed the status of investigations on color centers in alkali halide crystals at that time by classifying color centers into various kinds of defects, that is, vacancies, interstitials, impurities, and their complexes.

In crystalline semiconductors, defects have been investigated on irradiated crystals, for example, radiation-damaged crystalline silicon [Watkins, 2000]. Various types of defects such as single vacancy, divacancy, vacancy–hydrogen complex, and so on have been observed, using electron spin resonance (ESR) and electron–nuclear double resonance (ENDOR) spectroscopy. These spectroscopies provide powerful tools as well as optical means to elucidate the nature of defects. The g-values of these typical defects in crystalline silicon (c-Si) are shown in Fig. 3.9. In this book, however, this issue is not described and the results are only cited for comparison with amorphous and microcrystalline semiconductors. Even in crystals, for example, alkali halides and c-Si, it is worthy to note that various types of defects are present when specimens are either irradiated or chemically treated for alkali halides.

In amorphous semiconductors, we take hydrogenated amorphous silicon (a-Si:H) in Chapter 3 and amorphous chalcogenides (a-Ch) in Chapter 5. In a-Si:H, typical defects are normal silicon-dangling bonds and hydrogen-related dangling bonds (an analog of vacancy–hydrogen complex in c-Si), as will be mentioned in Chapter 3. The silicon-dangling bond is a threefold-coordinated silicon, that is, a broken bond. This broken bond in a-Si:H has been extensively investigated by means of ESR, ENDOR, electrically detected magnetic resonance (EDMR), and optically detected magnetic resonance (ODMR), as will be mentioned below and in Section 3.2.2. For EDMR and ODMR, see the next section. Very recently, silicon-dangling bonds in a-Si:H have been investigated, using multifrequency ESR spectroscopy along with density functional theory calculations [Pfanner *et al.*, 2011; Freysoldt *et al.*, 2012; Fehr *et al.*, 2011]. Hydrogenated micro-crystalline silicon is also treated in Chapter 4.

Identifications of defects in amorphous semiconductors are rather difficult in comparison with defects in crystalline semi-conductors, because amorphous semiconductors are a disordered system, so that the principal axes of symmetry of defects are randomly distributed over the amorphous network. In this case, comparison between observed ESR spectra and computer-simulated ones is useful. Further, ENDOR provides a powerful tool for determining the size and character of wave function of the unpaired electron responsible for ESR [Stutzmann and Biegelsen, 1986; Yokomichi and Morigaki, 1993].

1.3 Spin-Dependent Properties

Spin-dependent properties of semiconductors have been extensively investigated so far and are used for highly sensitive detection of magnetic resonance signals, that is, ESR signals, ENDOR signals, and so on. As spin-dependent properties, photoluminescence (PL), optical absorption, photoconductivity, dark conductivity, and so on are considered. In the following, we summarize spin-dependent properties of semiconductors.

Recombination, either radiative or nonradiative, occurs between an electron and a hole having different spin direction, antiparallel to each other. Triplet states (parallel spin states) have a long lifetime compared with its singlet states (antiparallel spin states). ODMR normally monitors the PL intensity (or emitted light-polarization in crystalline semiconductors), so that ESR of triplet states (either an electron or a hole in the parallel spin states) induces an increase in the PL intensity [Cavenett, 1981; Morigaki, 1982, 1984; Morigaki and Kondo, 1995]. When the spin-lattice relaxation time is longer than the lifetime of the electron–hole pair, the Zeeman levels of the electron–hole pair in the presence of magnetic field are unthermalized, that is, the population of anti-parallel spin states become smaller than those of parallel spin states [Kaplan *et al.*, 1978; Dunstan and Davies, 1979]. Then, one expects a big change in the PL intensity associated with ESR of either an electron or a hole of the electron–hole pair. Further, theory of ODMR has been developed by Movaghar *et al.* [1980] and Morigaki [1981]. Exchange interaction between spins has also been investigated by Murayama *et al.* [1975], using the ODMR technique. See also [Morigaki, 1999].

When an electron is excited from a bond, for example, an Si–Si bond in silicon, it is attracted by a hole left in the bond due to Coulomb interaction; such an electron–hole pair is called geminate electron–hole pair. On the contrary, distant electron–hole pairs such as those mentioned above are called nongeminate electron–hole pairs. For the geminate electron–hole pair, ESR of either an electron or a hole of the geminate electron–hole pair causes the PL intensity to be decreased, because ESR changes the antiparallel spin state of the geminate electron–hole pair to its parallel spin state; then the PL due to recombination of geminate electron–hole pairs decreases in intensity [Biegelsen *et al.*, 1978]. Thus, this change is opposite to that expected for nongeminate electron–hole pairs.

Spin-dependent recombination can be observed through ESR of recombination centers [Lepine, 1972; Lepine *et al.*, 1976; Solomon *et al.*, 1977; Solomon, 1979]. This phenomenon is generally used for observation of the ESR signal of recombination center, monitoring photocurrents under illumination, that is, this is sometimes called EDMR [Lips *et al.*, 1996]. See also [Morigaki, 1999]. At low temperatures, neutral impurity scattering affects photoconductivity, for example, in lightly doped crystalline silicon, so that ESR of impurities changes their spin polarization so as to affect the photoconductivity [Honig, 1966; Schmidt and Solomon, 1966]. This effect has been observed by Schmidt and Solomon [1966] and Maxwell and Honig [1966]. Such spin-dependent effect has been reviewed by Solomon [1972].

Spin-dependent phenomena were first observed in electrical conduction of doped semiconductors, that is, n-type InSb by Guéron and Solomon [1965] and reduced rutile by Hirose *et al.* [1966]. After then, spin-dependent conductivity has been extensively investigated in n-type silicon [Toyoda and Hayashi, 1970, 1971; Toyotomi and Morigaki, 1970; Morigaki and Toyotomi, 1971; Morigaki and Onda, 1974; Toyotomi, 1974; Morigaki *et al.*, 1975; Kishimoto *et al.*, 1977] and n-type germanium [Morigaki and Onda, 1972]. Spin-dependent hopping conduction has been observed and discussed in amorphous silicon by Kishimoto *et al.* [1981] and also discussed in doped c-Si by Kamimura and Mott [1976].

Spin-dependent effects on the photoinduced absorption (PA) in a-Si:H [Hirabayashi and Morigaki, 1983; Schultz *et al.*, 1997], on the a-Si:H pin solar cells [Homewood *et al.*, 1983; Lips and Fuhs, 1991; Fuhs and Lips, 1993], on the a-Si:H Schottky barrier diodes [Lerner *et al.*, 1998], and on the c-Si pin diode [Solomon, 1976; Solomon and Bhatnagar, 1992] have also been observed so far. See also [Stutzmann *et al.*, 2000].

Recently, the observation of Rabi oscillation has been made in pulsed ODMR and EDMR experiments, from which information about dipolar interaction between two carriers or a carrier and a recombination center may be obtained [Boehme and Lips, 2003; Lips *et al.*, 2005; Fuhs, 2008].

Chapter 2

Crystalline Semiconductors

2.1 Introduction

This section deals with light-induced defects in crystalline semiconductors, particularly in GaAs and related materials. Defects are created by optical excitation through nonradiative recombination between optically generated electrons and holes, but, in these materials, photoelectrons or holes are replaced by injected electrons or holes for their roles in the light-induced defect creation, because the defect creation is also associated with injection of electrons or holes, for example, in light-emitted diodes. In Section 2.2, experimental evidence and theories of recombination-enhanced defect reaction are described with an emphasis on GaAs and related materials [Lang, 1992; Mooney, 1992]. In Section 2.3, persistent photoconductivity and the DX centers are treated on GaAs and related materials. In Section 2.4, hydrogenated polycrystalline silicon is also treated in relation with a-Si:H.

Light-Induced Defects in Semiconductors
Kazuo Morigaki, Harumi Hikita, and Chisato Ogihara
Copyright © 2015 Pan Stanford Publishing Pte. Ltd.
ISBN 978-981-4411-48-6 (Hardcover), 978-981-4411-49-3 (eBook)
www.panstanford.com

2.2 Recombination-Enhanced Defect Reaction

2.2.1 Introduction

The annealing rate at room temperature of radiation-induced defects in GaAs has been found by Lang and Kimerling [1974] to be enhanced as a result of nonradiative recombination at defects, which is increased by a factor of 10^6 when minority carriers are injected into a sample. This recombination energy is utilized to defect diffusion, which turns out to annealing of defects. Such so-called recombination-enhanced defect reaction (REDR) has been established by Lang and Kimerling [1974] from their finding that the annealing rate of defects is connected to the electron–hole recombination at the defect. The radiation-induced defect lies at 0.31 eV below the conduction band edge. After then, Weeks *et al.* [1975] developed a theory of REDR on the basis of the following mechanism: energy liberated upon nonradiative electron or hole capture is converted largely into vibration energy that is initially localized in the vicinity of the defect. This vibration energy can be utilized to promote defect reaction such as diffusion. They employed the Rice–Ramsperger–Kassel theory of unimolecular reactions, which deals with a defect and its immediate neighborhood, that is, an almost isolated defect molecule embedded in the host lattice. Further, Sumi [1984] developed a dynamical theory of REDR. In the following, we present the experimental result by Lang and Kimerling [1974], and then we mention about the above two theories.

2.2.2 Experimental Evidence

A deep level (called the E3 level) at 0.31 eV below the conduction band edge is formed by 1 MeV electron irradiation. It captures an electron, and then a hole introduced by the injection is captured, following recombination between the electron and the hole. This causes the defects to be annealed out as a result of REDR. The temperature dependences of the annealing rate for the E3 level in GaAs are shown in Fig. 2.1 without (thermal) and with (saturated) hole injection. In Fig. 2.1, the case with hole injection corresponds to the saturated annealing rate. The activation energy for annealing is evaluated from the slope of the temperature dependence in

Fig. 2.1. Associated with the hole injection, the activation energy is greatly reduced from 1.4 ± 0.15 to 0.34 ± 0.03 eV. This is an evidence for REDR. This result has been theoretically discussed by Weeks *et al.* [1975] and Sumi [1984].

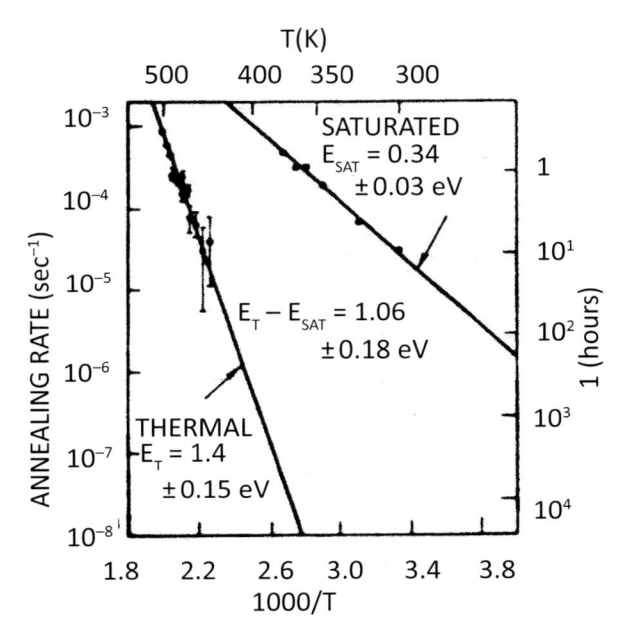

Figure 2.1 Temperature dependence of annealing rate without (thermal) and with (saturated) hole injection. Activation energies are shown. [Reproduced from Kimerling, *Solid State Electron.*, **21**, 1391 (1978) by permission of Elsevier.]

2.2.3 Theory

In this section, we deal with two theories of REDR by Weeks *et al.* [1975] and Sumi [1984]. First, we summarize a theory of Weeks *et al.* [1975]. The energy diagram involved in their theory is shown in Fig. 2.2, in which a deep level is illustrated. They take a model of the defect molecule that represents a defect and its immediate neighborhood. Using this model as well as the Rice–Ramsperger–Kassel theory of unimolecular reactions, they developed a theory of REDR as follows: Associated with nonradiative capture of either electron or hole at the deep level in Fig. 2.2, the vibration energy may be supplied to the lattice through strong electron–phonon

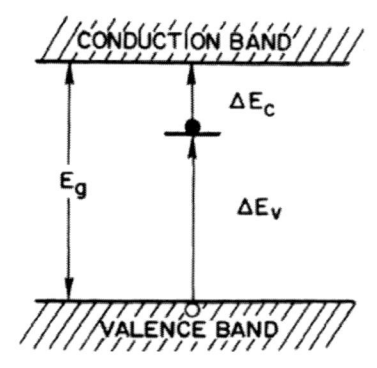

Figure 2.2   Energy position of an occupied deep state with associated electron and hole transitions. [Reproduced from Weeks *et al.*, *Phys. Rev. B*, **12**, 3286 (1975), by permission of American Physical Society.]

interaction. In this theory, $n(E)$ is defined so that $n(E)\,dE$ is the number of defect molecules possessing the vibration (internal) energy between E and $E + dE$. The two rates are also defined, one of them, $r(E, E')$, is related to deactivation (loss of energy for the lattice) when $E > E'$ and thermal activation when $E < E'$, and the other $r_3(E, E')$, that is, the capture rate of an electron or a hole accompanied by the energy transfer ΔE to the lattice ($\Delta E \cong \Delta E_C$ or ΔE_V) in Fig. 2.2 and this is also defined in terms of the recombination rate R as follows:

$$r_3(E, E') = R\delta(E - E' - \Delta E), \tag{2.1}$$

The internal flow of energy within the defect molecule is defined by $k_2(E)$, which is the rate from the accepting coordinate to the reaction coordinate of vibration energy, that is, $k_2(E) = 0$ for $E < E^*$ and $k_2(E) \neq 0$ for $E \geq E^*$, where E^* is the minimum energy necessary for the internal flow. Weeks *et al.* construct the rate equation of $n(E)$, and they use the steady-state value of $n(E)$ because $n(E)$ reaches its steady state after an initial transient period. The total rate of diffusion step necessary for defect reaction, for example, recombination-enhanced annealing of defects is given by

$$I_d = \int_{E^*}^{\infty} k_2(E)n(E)dE \equiv Nk_d, \tag{2.2}$$

where N and k_d are the total number of defects and the diffusion rate constant, respectively. In the integral of Eq. (2.2), the recombination events are sufficiently so rare that $n(E)$ is replaced by the thermal equibrium value, $n_{eq}(E)$. The thermal equilibrium fraction of defects $f(E)dE$ with energy between E and $E + dE$ is defined by

$$f(E) = n_{eq}(E)/N. \tag{2.3}$$

The diffusion rate constant, k_d, is the sum of two terms;

$$k_d = k_d \text{ (thermal)} + k_d \text{(enhanced)}, \tag{2.4}$$

where k_d (thermal) is the diffusion rate from ordinary thermal processes and k_d (enhanced) is the extra diffusion caused by the capture process.

As $f(E)$ is calculated from the use of the Rice–Ramsperger–Kassel theory of unimolecular reaction, Weeks *et al.* obtain the final form of k_d (thermal) and k_d(enhanced) as follows:

$$k_d(\text{thermal}) = k^* \int_{E^*}^{\infty} \left(\frac{E-E^*}{E}\right)^{S-1} \frac{1}{(S-1)!}\left(\frac{E}{kT}\right)^{S-1} \frac{1}{kT} e^{-E/kT} dE$$

$$= k^* \, e^{-E^*/kT}, \tag{2.5}$$

$$k_d(\text{enhanced}) = R \int_{E^*}^{\infty} \frac{k^*}{k_{-1}(E)} \left(\frac{E-E^*}{E}\right)^{S-1} \frac{1}{(S-1)!}\left(\frac{E-\Delta E}{kT}\right)^{S-1} \frac{1}{kT} e^{-(E-\Delta E)/kT} dE \tag{2.6}$$

$$k_2 = k^* \left(\frac{E-E^*}{E}\right)^{S-1}, \tag{2.7}$$

$$k_{-1}(E) = \int_0^{\infty} r\,(E, E') dE', \tag{2.8}$$

where S is the number of loosely coupled oscillators with the same vibrational frequency, and k^* is the prefactor. As seen in Eq. (2.5), E^* is identified as the activation energy for diffusion.

For comparison with the experimental result for radiation-induced defects in GaAs, Weeks *et al.* obtain the following approximate expression adequate for the experimental results:

$$k_d(\text{enhanced}) = R \frac{k^*}{k_{-1}(E)}^{S-1} \left(\frac{E^* - \Delta E}{E^*} \right)^{S-1} e^{-(k^* - \Delta E)/kT}, \qquad (2.9)$$

$$R = R_0 e^{-E_R/kT}, \qquad (2.10)$$

The annealing rate k_a is related to the diffusion step rate as follows:

$$k_a = \frac{k_d}{N_j} = K_a e^{-E_e/kT}, \qquad (2.11)$$

where N_j is the average number of diffusion steps.
The prefactor K_a is given by

$$k_a = R_0 \frac{k^*/N_j}{k_{-1}(E)} \left(\frac{E^* - \Delta E}{E^*} \right)^{S-1}, \qquad (2.12)$$

and the enhanced activation energy is given by

$$E_e = E^* - \Delta E + E_R \qquad (2.13)$$

The calculated results are compared with the experimental ones obtained for GaAs. For this sample, E_g = 1.4 eV, ΔE_C = 0.31 eV, and ΔE_V = 1.09 eV. Assuming $\Delta E \cong \Delta E_V$ = 1.09 eV for hole capture and using experimental values for E^* from Eq. (2.5) and E_R from Eq. (2.10), we obtain

$$E_e = (1.4 \pm 0.15) - 1.09 + 0.1 = (0.41 \pm 0.15) \text{ eV}, \qquad (2.14)$$

This agrees with the experimental result E_e = 0.34 eV.
The prefactor is

$$K_e = 5 \times 10^8 [10^{11}/k_{-1}(E^*)](0.17)^{S-1} \qquad (2.15)$$

For Eq. (2.15), we use Eq. (2.13), in which $E^* - \Delta E = E_e - E_k = 0.17$ eV, where E_e = 0.34 eV, E_k = 0.1 eV, and E^* = 1.4 eV.

Two unknown quantities are included in Eq. (2.15), that is, $k_{-1}(E^*)$, the average rate of equilibration with the lattice, and S, the number of effective oscillators in the defect molecule.

For two choices of $k_{-1} = 10^{11}$ and 10^{12} s^{-1}, then we fit the data with $S = 10$ and $S = 8$, respectively. Weeks *et al.* conclude that the number of effective oscillators S is not very sensitive to the choice of k_{-1}, and the predicted values of S are reasonable for the number of localized modes around a point defect.

In the following, we deal with a dynamical theory of REDR by Sumi who takes into account the efficiency of vibrational energy from the accepting coordinate (mode) to the reaction coordinate (mode), using the configuration coordinate model. The configuration coordinate model is useful for understanding the basic process of nonradiative capture of free carriers by a deep level, as illustrated in Fig. 2.3. A free carrier is excited from the bottom of continuous levels just to the crossing point, and then it goes down to the deep level by emitting multiphonons, as shown in Fig. 2.3. Figure 2.4 illustrates several processes from the free carrier's capture to the defect reaction. The accepting coordinate Q_p and the reaction coordinate Q_R are shown in a plane, that is, the cross-section of the

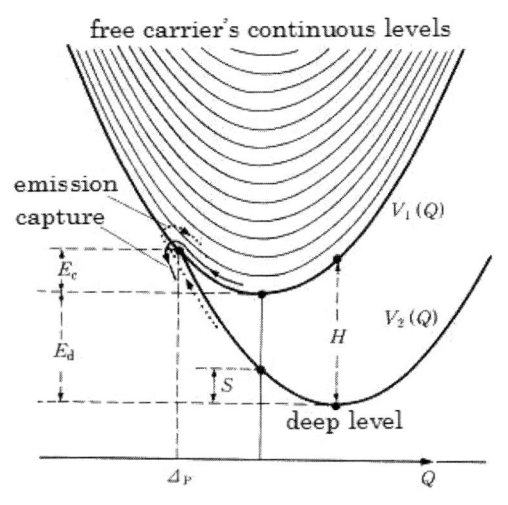

Figure 2.3 Schematic diagram of the configuration coordinate model, in which free carriers are captured by deep levels and captured carriers are reemitted. [Reproduced from Sumi, *Kotai Butsuri* (*Solid State Physics*), **23**, 221 (1988) by permission of AGNE Gijutsu Center.]

adiabatic potential energy surface (two-dimensional potentials) obtained after free carrier's capture including the two coordinates. The cross-section of this potential plane along the coordinate Q_p corresponds to a potential $V_2(Q)$ in Fig. 2.3. The free carrier's capture occurs along a line of $Q_p = \Delta_p$. After free carrier's capture, multiphonon emission occurs along the accepting coordinate Q_p. The dissipated energy associated with multiphonon emission is defined by E_p that measures from a point of O that has minimum energy (see Fig. 2.4). The defect reaction occurs when the value of Q_R crosses over a line of $Q_R = \Delta_R$. The minimum energy point along a line of $Q_R = \Delta_R$ is R, whose energy is E_A measured from a point of O. E_A corresponds to the activation energy for the defect reaction in the thermal equilibrium of the phonon system after free carrier's capture. In REDR, the activation energy of the defect reaction may be reduced from E_A with E_p, that is the dissipated energy (vibrational energy) when multiphonon emission occurs. In Fig. 2.4, lattice relaxation occurs along the coordinate Q_p, and then defect reaction occurs along the coordinate Q_R. Two processes are connected as a dotted curve, as shown in Fig. 2.4. The starting point is K in Fig. 2.4, which is called the phonon-kick point. The dotted curve corresponds to the trajectory with the lowest energy connecting two points K and R. The energy at K is higher than that at P, so that the energy is written by $E_p + E_H$. E_H is energy necessary for reaching at R, whose additional energy should be supplied from the phonon system, corresponding to the thermal activation energy of the quantum yield for the defect reaction. The efficiency of the dissipated energy to the defect reaction depends on the direction cosine g between two directions of the coordinates Q_p and Q_R in Sumi's theory. Sumi [1984] has shown that the thermal activation energy of the quantum yield of the defect reactions is reduced, as shown in Fig. 2.5. For recombination-enhanced annealing of the E_3 level in GaAs, in the case of low cosine injection level, he obtained $g = 0.65$, using various values for the parameters involved in the calculated result, that is, $E_g = 1.4$ eV, $E_A = 1.34$ eV, $E_k = 0.43$ eV, $E_p = 1.24$ eV, $E_{1c} = 0.33$ eV, and so on. The defect molecule model assumes 100% efficiency for transfer of vibrational energy to the defect reaction, but according to Sumi's theory, this is not true, as shown above.

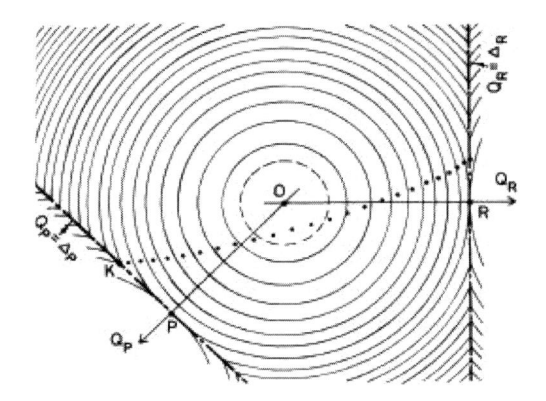

Figure 2.4 Adiabatic potential-energy surface after free carrier's capture in a plane with the energy-accepting coordinate Q_P and the reaction coordinate Q_R. The dotted curve shows a trajectory of the phonon system starting from a line $Q_P = \Delta_P$, where carrier's capture occurs, and reaching a line $Q_R = \Delta_R$, where defect reaction occurs. [Reproduced from Sumi, *Phys. Rev. B*, **29**, 4616 (1984) by permission of American Physical Society.]

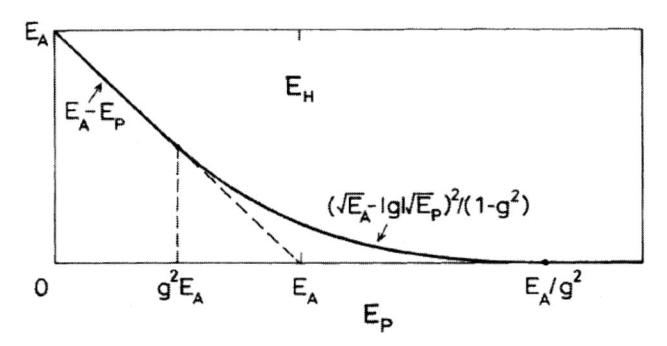

Figure 2.5 Thermal activation energy E_H of the quantum yield of REDR as a function of phonon-kick energy E_P. [Reproduced from Sumi, *Phys. Rev. B*, **29**, 4616 (1984) by permission of American Physical Society.]

The E3 level has been identified to be a Ga vacancy. The vacancy is annealed out when it encounters an interstitial Ga atom after its diffusional jump. Some of the Ga vacancies are trapped by other types of defects before reaching an interstitial Ga atom.

2.2.4 Persistent Photoconductivity and the DX Centers

The persistent photoconductivity is one of the striking phenomena observed for alloy materials such as n-type $Ga_{1-x}Al_xAs$, for example, Te-doped $Ga_{1-x}Al_xAs$ ($0.25 \leq x \leq 0.7$) [Nelson, 1977]. The increase in conductivity caused by sub-bandgap illumination can last for long times at low temperatures, as shown in Fig. 2.6.

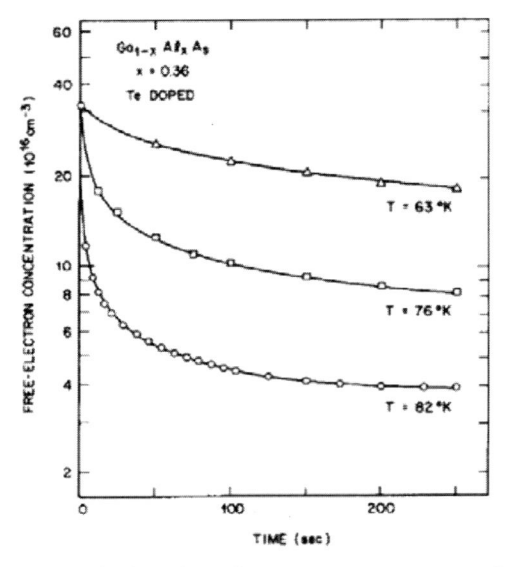

Figure 2.6   Decay of the free-electron concentration (inverse Hall coefficient: $1/eR_H$) after photoexcitation for three different temperatures. In each case, $1/eR_H$ was saturated during a long excitation period. [Reproduced from Nelson, *Appl. Phys. Lett.*, **31**, 351 (1977) by permission of American Institute of Physics.]

The inverse Hall coefficient $1/eR_H$ corresponding to carrier concentration has been measured as a function of the inverse of temperature in the dark and under illumination (the saturated level), as shown in Fig. 2.7. From these figures, we find that the high electron concentration persists with a long decay time after illumination is turned off. Such a long decay time indicates that the capture cross-section for the photoexcited carriers is less than 10^{-30} cm^2, which is six orders of magnitudes smaller than the normal capture cross-section due to ionized or neutral impurity

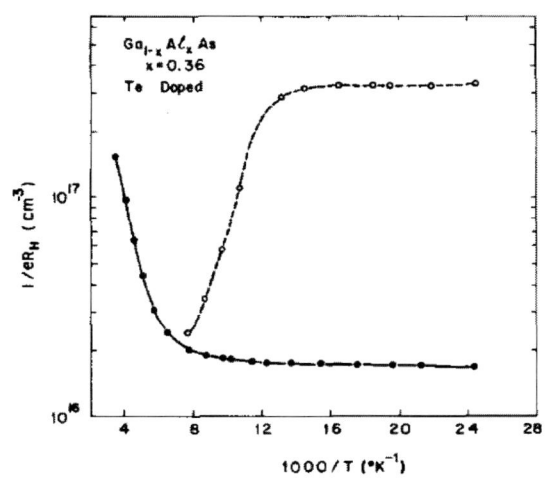

Figure 2.7 Temperature dependence of $1/eR_H$. The solid curve represents data taken with the sample in the dark, while the dashed curve does data taken while the electron population is saturated under photoexcitation. [Reproduced from Nelson, *Appl. Phys. Lett.*, **31**, 351 (1977) by permission of American Institute of Physics.]

scattering at low temperatures in crystalline semiconductors. Furthermore, the capture process is thermally saturated at a temperature below 100 K, and the thermal barrier height has been estimated from the decay curve of the free-electron concentration shown in Fig. 2.6 to be 0.18 eV in $Ga_{1-x}Al_xAs$ ($x = 0.36$) [Nelson, 1977]. However, the center responsible for the electron capture has a large Stokes shift. The photoionization energy is 1.1 eV for this material. Examples of photoionization lineshape in Te, Sn, and Si-doped n-type $Ga_{1-x}Al_xAs$ are shown in Fig. 2.8, which have been measured on Schottky barrier samples at 50 K, using photocapacitance. The configuration coordinate model for the center is shown in Fig. 2.9. As one of models for the center, the DX center has been proposed, which is the donor atom associated with an unknown defect (X), because the above observation is strongly connected with the shallow donor concentration. In Fig. 2.9, curve C and V correspond to the total system energy of an unoccupied defect with a delocalized electron in either the conduction or valence band, respectively. Curve D corresponds to the vibrations of an occupied defect. The thermal barriers to electron capture and to electron emission are designated as E_{ec} and E_{ee}, respectively. In the figure, $E_{ec} \cong 0.2$ eV and $E_{ee} \cong 0.3$ eV are taken from the following measurements. Electron emission and electron capture rates have been measured from junction capacitance spectroscopy by Lang and Logan [1977]. The results are shown in Fig. 2.10. In the figure, the photoconductivity data are taken from Nelson [1977]. He concluded that the center observed at high concentration of donors in Te, Se, and Sn-doped $Ga_{1-x}Al_xAs$ ($0.25 \leq x \leq 0.70$) has a much larger optical (1.1 eV) than thermal (0.12 eV) ionization energy, so that the lattice relaxation shown in Fig. 2.9 creates an energy barrier to electronic capture that produces a persistent photoconductivity at low temperatures.

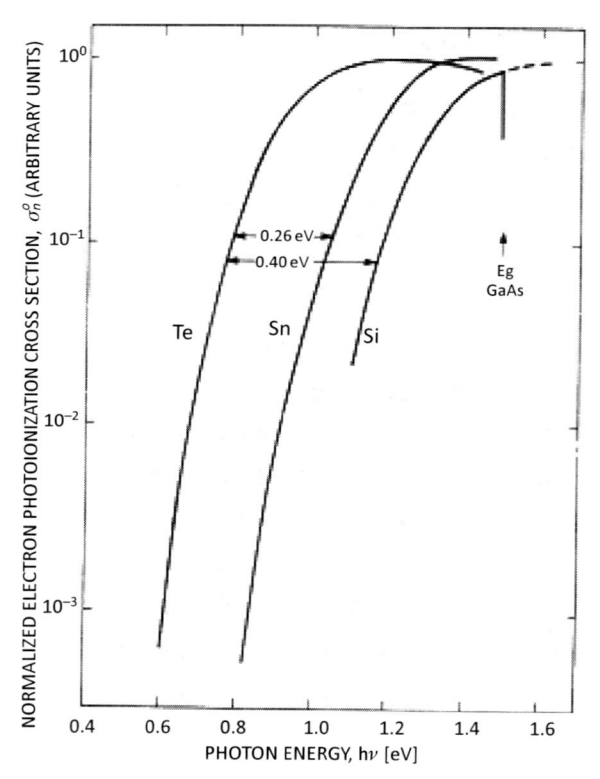

Figure 2.8 Photoionization lineshape of Te-, Sn-, and Si-related DX centers in n-type $Al_{0.4}Ga_{0.6}As$. [Reproduced from Lang and Logan, *Physics of Semiconductors*, 1978, Conf. Series No. 43, 433, by permission of Institute of Physics, UK.]

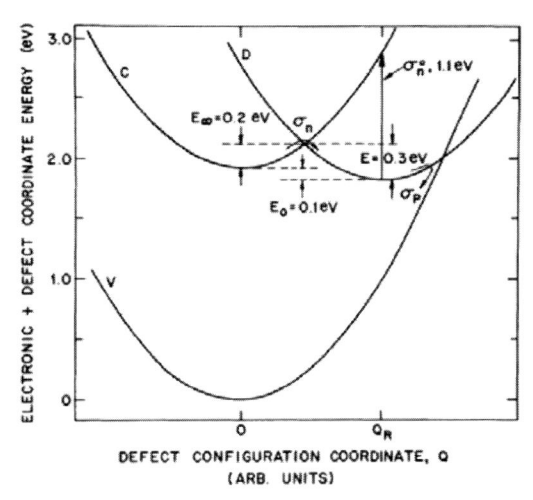

Figure 2.9 Configuration coordinate model for the donor-defect center in $Al_xGa_{1-x}As(Te)$. [Reproduced from Lang and Logan, *Phys. Rev. Lett.*, **39**, 635 (1977) by permission of American Physical Society.]

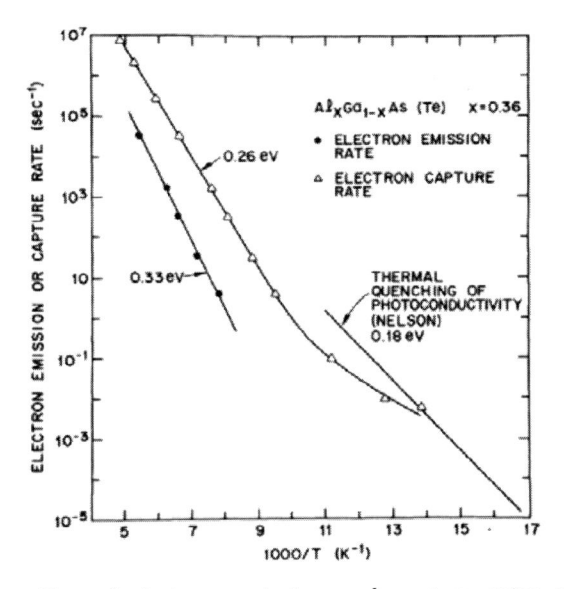

Figure 2.10 Plot of electron emission and capture rates *vs.* inverse temperature for a typical sample of $Al_xGa_{1-x}As(Te)$. [Reproduced from Lang and Logan, *Phys. Rev. Lett.*, **39**, 635 (1977) by permission of American Physical Society.]

The microscopic model for the DX center is described as follows: A model proposed by Chadi and Chang [1989] is called distorted DX center, that is, a bond between Si atom and neighboring As atom is broken, so that the Si atom is threefold-coordinated with neighboring As atoms, as shown in Fig. 2.11(b). An Si atom is substituted to a Ge site, as shown in Fig. 2.11(a), while it moves toward an adjacent interstitial site along <111> direction, as shown in Fig. 2.11(b). The distorted configuration shown in Fig. 2.11(b) corresponds to the DX center proposed by Chadi and Chang [1989]. For this configuration, the two electron state (the negatively charged state) is stable, so that it is nonmagnetic. This is consistent with no observation of the ESR signal from the DX center. For alloys such as $Ga_{1-x}Al_xAs$, an Al atom occupies a Ga site adjacent to the Si atom with its number 0, 1, 2, and 3, as shown in Fig. 2.12(a) and 2.12(b). Thus, four deep levels corresponding to 0, 1, 2, and 3 are formed. This gives four peaks in the DLTS spectra at maximum and also gives rise to broadening of the peak.

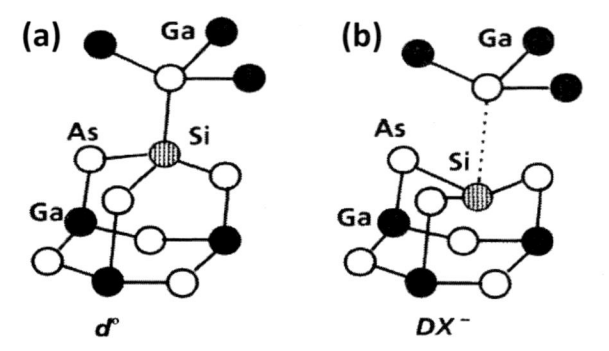

Figure 2.11 Schematic representation of the normal substitutional sites and the broken-bond configuration in Si-doped GaAs. [Reproduced from Chadi and Chang, *Phys. Rev. B*, **39**, 10063 (1989) by permission of American Physical Society.]

The fourfold-coordinated Si donor forms a shallow level, whereas the threefold-coordinated Si donor being the distorted configuration forms a deep level. These two levels correspond to the C and D curve of the total energy shown in Fig. 2.9. The VI atom such as S, Se, and Te occupies the As site having neighboring three Ga sites, as shown in Fig. 2.13. An As–Ga bond shown in

Fig. 2.13(a) is broken, so that a distorted configuration forms, as shown in Fig. 2.13(b).

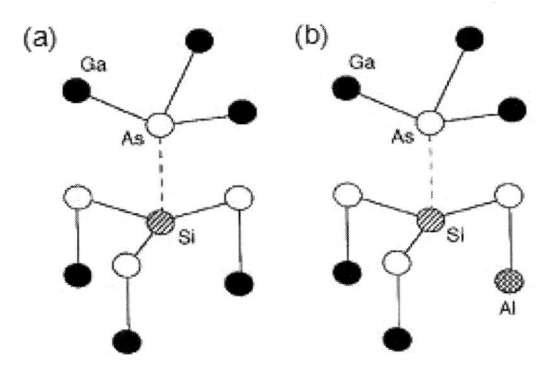

Figure 2.12 Schematic representation of the broken-bond configuration in Si-doped GaAs and AlGaAs. [Reproduced from Redfield and Bube, Photoinduced Defects in Semiconductors (1996) by permission of Cambridge University Press.]

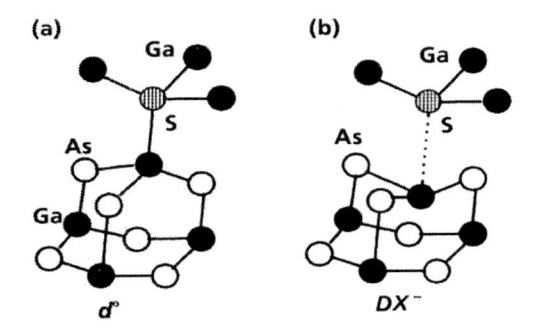

Figure 2.13 Schematic representation of the normal substitutional sites and the broken-bond configuration in S-doped GaAs. [Reproduced from Chadi and Chang, *Phys. Rev. B*, **39**, 10063 (1989) by permission of American Physical Society.]

2.3 Hydrogenated Polycrystalline Silicon

Polycrystalline silicon (poly-Si) contains the grain boundary in which Si-dangling bonds are observed, for example, at the density of 2.0×10^{19} cm^{-3} [Nickel *et al.*, 1993]. The poly-Si, for example, consists of columnar grains extending from the substrate to the sample surface with diameters of ~100 Å. Hydrogen incorporation is performed by an optically isolated remote hydrogen plasma, with which Si-dangling bonds are passivated, for example, until the density of 2.2×10^{16} cm^{-3} [Nickel *et al.*, 1993]. Such hydrogenated poly-Si (poly-Si:H) exhibits light-induced defect creation such as a-Si:H, for example, water-filtered light from a xenon-arc lamp at 7 W/cm^2 for 15.5 h at 80°C creates Si-dangling bonds with the density of 4.4×10^{16} cm^{-3} [Nickel *et al.*, 1993]. The repeated light-induced creation and annealing were performed like pm-Si:H and a-Si:H such as will be mentioned in Section 3.5.8. Figure 2.14 shows the result of the repeated light-soaking (LS) –160°C annealing cycle on the spin density, N_S, of poly-Si:H [Nickel *et al.*, 1993]. A slight decrease in N_S taken just after LS with the repeated LS-annealing cycles is seen, while N_S taken after annealing is increased with the repeated LS-annealing cycles. The observation that the efficiency of light-induced defect creation is slightly

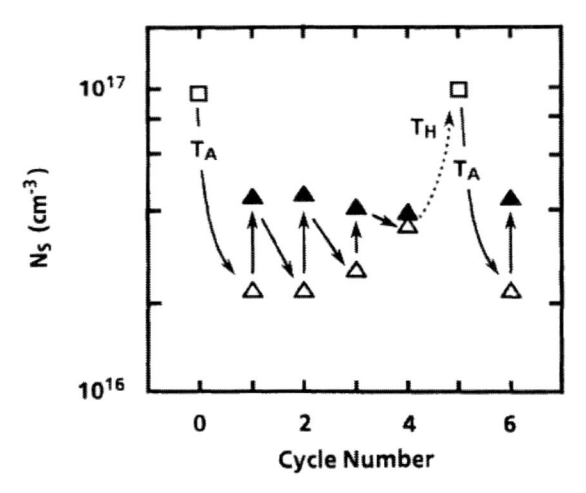

Figure 2.14 Spin density N_S of hydrogenated polycrystalline silicon films for several defect creation and annealing cycles. The open and closed triangles represent the anneal state and the light-soaked state, respectively. The degradation was carried out with 7 W/cm^2 white light for 15.5 h. The squares were obtained after hydrogenation at T_H = 350°C. Anneal: T_A = 160°C for 15 h. [Reproduced from Nickel *et al.*, *Phys. Rev. Lett.*, **71**, 2733 (1993) by permission of American Physical Society.]

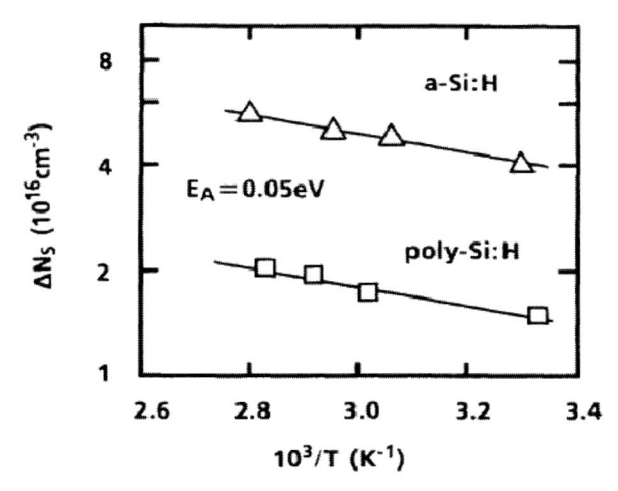

Figure 2.15 Temperature dependence of the change of the spin density ΔN_S in poly-Si:H and a-Si:H after prolonged illumination with 7 W/cm^2 white light for 15.5 h. [Reproduced from Nickel *et al.*, *Phys. Rev. Lett.*, **71**, 2733 (1993) by permission of American Physical Society.]

decreased is like that for pm-Si:H, as will be mentioned in Section 3.5.8. However, the annealing efficiency becomes worse with the repeated LS-annealing cycles. Further hydrogenation at 350°C increases N_S to its initial value before first annealing, as seen in Fig. 2.14. After then, annealing makes N_S come back to its initial annealed value, and then LS gives rise to light-induced defect creation such as the first stage of the repeated LS-annealing cycle. Nickel *et al.* [1993] suggested from these results that hydrogen is directly involved in the creation and annealing of Si-dangling bonds. The reason of why the efficiency of light-induced defect creation in poly-Si:H is slightly decreased with the repeated LS-annealing cycles is in part that strongly bonded hydrogen is formed, including large H platelets or cluster or molecules in voids, because such hydrogen cannot participate in the creation and annealing of Si-dangling bonds. A comparison in the behavior of LS-annealing cycles with those of a-Si:H and pm-Si:H will be mentioned in Section 3.5.8. Nickel *et al.* [1993] measured the light-induced defect density ΔN_S as a function of illumination temperature, as shown in Fig. 2.15, in which the case of a-Si:H is also included. The activation energy is 50 meV, which is the same as that of a-Si:H. From a fact that light-induced defect creation is observed in poly-Si:H, similarly to a-Si:H, it may be concluded that hydrogen plays an important role in such phenomena.

Chapter 3

Hydrogenated Amorphous Silicon

3.1 Introduction

Amorphous semiconductors are a typical example of the disordered system. For reviews, see Mott and Davis [1979], Morigaki [999], Singh and Shimakawa [2003], Morigaki and Ogihara [2006]. The a-Si:H is one of amorphous materials having extensively investigated so far from the viewpoints of physics and applications such as solar cells and thin-film transistors. In this chapter, we deal with a-Si: H, particularly on the light-induced defects in this material. Thus, first the a-Si:H material is briefly reviewed with an emphasis on the structure in this section [Mott and Davis, 1979; Morigaki, 1999; Singh and Shimakawa; 2003; Morigaki and Ogihara, 2006; Street, 1991; Matsuda, 2006]. The Si atoms are fourfold-coordinated with four neighboring Si atoms, but the long-range order is not held in the structure, that is, so-called amorphous network. Furthermore, hydrogen is incorporated into the network in the form of SiH, SiH_2, SiH_3, and H_2. In high-quality a-Si:H, the principal configuration of hydrogen is SiH, while, in low-quality a-Si:H, SiH_2 and SiH_3 are also involved as well as SiH. Actual a-Si:H samples contain Si-dangling bonds, that is, neutral threefold-coordinated Si atoms as

Light-Induced Defects in Semiconductors
Kazuo Morigaki, Harumi Hikita, and Chisato Ogihara
Copyright © 2015 Pan Stanford Publishing Pte. Ltd.
ISBN 978-981-4411-48-6 (Hardcover), 978-981-4411-49-3 (eBook)
www.panstanford.com

well as charged Si-dangling bonds, for example, with the density of 10^{15}–10^{16} cm^{-3} for high-quality samples, and 10^{17}–10^{18} cm^{-3} for low-quality samples. The hydrogen content depends on the preparation method and preparation condition [Street, 1991; Yamaguchi and Morigaki, 1999]. The a-Si:H samples are often prepared by plasma-enhanced chemical vapor deposition (PECVD). For samples prepared by PECVD, high-quality samples contain hydrogen with [H] $\cong 10$ at.%, while low-quality samples contain hydrogen with [H] $\cong 20$–30 at.%. The a-Si:H samples also contain voids with a dimension of 30–100 Å, around which Si-dangling bonds and hydrogen exist. The structure of amorphous network of Si is normally represented by a continuous random network model, in which the bond length and the bond angle have spatial fluctuations within ±1% and ±10%, respectively, and the dihedral angle is distributed over the network.

Silane gas diluted with hydrogen gas is often used for preparation of a-Si:H. The silane fraction (gas-dilution ratio), γ, is defined by $\gamma = [SiH_4]/([SiH_4] + [H_2]$. However, if γ is smaller than a critical value of γ, γ_c, microcrystalline grains are mixed into the matrix of amorphous silicon, that is, the film consists of two phases of microcrystalline and amorphous phases. This is called hydrogenated microcrystalline silicon (μc-Si:H). The value of γ_c depends upon preparation conditions, for example, substrate temperature, T_S. For PECVD, γ_c is 3.5% for $T_S = 100°C$ [Yamaguchi and Morigaki, 1993], while, for hot-wire CVD (HWCVD), γ_c is 8% for $T_S = 230°C$ [Niikura *et al.*, 2011]. The crystalline volume fraction is estimated from Raman scattering at 514 or 633 nm laser radiation via the TO-phonon peak. The X-ray diffraction pattern is a powerful means for examining the preferential orientation and the microcrystalline grain size. The μc-Si:H films are treated in Chapter 4. For structural and other properties, see also [Street, 1991], [Vetterl *et al.*, 2000], [Niikura *et al.*, 2011].

At high pressure of mixed gas, for example, between 1200 and 2200 mTorr and rf power of 22 W, a new type of films are prepared [Roca i Cabarrocas, 2000; Roca i Cabarrocas *et al.*, 2002; Roca i Cabarrocas *et al.*, 2004], which is called hydrogenated polymorphous silicon (pm-Si:H). Below 1200 mTorr, μc-Si:H films are prepared. The measurements of Raman spectra [Butté *et al.*, 2000] and transmission electron microscopy [Fontenberta *et al.*, 2000] suggest that pm-Si:H consists of a relaxed amorphous network in which a low fraction of small crystallites with 2–4 nm in size is embedded. The crystallite concentration was roughly estimated to be of the order of 10^{18} cm^{-3} corresponding to a crystallite fraction of about 2% [Roca i Cabarrocas *et al.*, 2002]. Hydrogen is incorporated into films in pm-Si:H more than in a-Si:H deposited at the same temperature and deposition rate, that os, for an example, standard a-Si:H films deposited at 150°C contain bonded hydrogen with about 12 at.%, while pm-Si:H films contain hydrogen with about 18 at.% [Roca i Cabarrocas *et al.*, 2002]. The following hydrogen incorporation scheme has been obtained from infrared absorption measurement [Roca i Cabarrocas *et al.*, 2002]: Diluted Si–H bonds, SiH$_2$ bonds, and/or Si–H bonds on the surface of internal cavities, and Si–H bonds on the surface of crystallites. In a-Si:H, with increasing hydrogen content from 12 at.% to 20 at.%, the film density decreases from 2.30 to 2.10 g/cm^3, while in pm-Si:H, it is kept between 2.20 and 2.25 g/cm^3 in the range of hydrogen content of 12–23 at.%.

3.2   Electronic States

3.2.1   Introduction

The electronic states in amorphous semiconductors can be generally treated in terms of tight-binding approximation. As shown in Fig. 3.1, the structural unit consists of a cluster of tetrahedral coordination with neighboring Si atoms in a-Si. The bonding state forms the valence band, while the antibonding state forms the conduction band. The binding energy of the Si–Si bond varies with site to site because bonding length, bonding angle, and dihedral angle have spatial fluctuation. As a result, the energy separation between the bonding state and the antibonding state varies spatially in the amorphous network. As the energy separation corresponds to the energy gap of the conduction and valence bands, such a spatial fluctuation forms the tail states in each band. These tail states are of localized nature in the sense of Anderson localization. The boundary between the localized states and the delocalized states is called the mobility edge and the energy separation between two boundaries in respective bands is called the mobility gap. This concept is called the Mott-CFO (Cohen, Fritzsche and Ovsinsky) model [Mott, 1970], [Cohen *et al.*, 1969]. See also [Taylor, 2006].

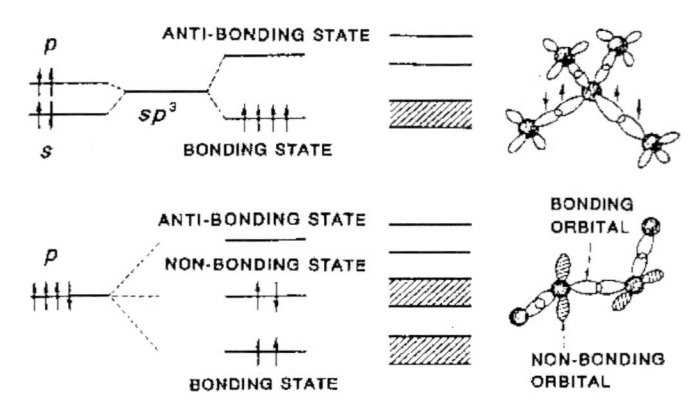

Figure 3.1   Schematic diagram of the energy levels of atomic orbitals, hybridized orbitals, and bands for (a) a tetrahedrally bonded semiconductor, for example, a-Si, and for (b) selenium.

The detailed information of the band tail and the band itself can be generally obtained through various experimental means such as optical absorption, photoemission spectroscopy, luminescence, transport, and magnetic resonance. Actually, amorphous semiconductors prepared by various means have no ideal amorphous network, but they contain structural defects and distorted network. Here, we deal with the tail states and the structural defects in a-Si and a-Si:H, particularly experimental information obtained from microscopic means such as magnetic resonance (NMR, ESR, ENDOR), optically detected magnetic resonance (ODMR), and so on.

The atomic scale structure investigations of tetrahedrally bonded amorphous semiconductors using the tight-binding molecular dynamics simulations and reverse Monte Carlo structural modeling method predict the existence of threefold and fourfold rings that are new types of defect (triangles and squares) [Kugler, 2012].

3.2.2 Band Tails and Structural Defects

The density of states (DOS) of the band tail has generally the following shape,

$$f(E) = f(E_0)\exp\{-(E - E_0)/\Delta E\} \tag{3.1}$$

From various types of measurements in a-Si:H, the values of ΔE have been obtained to be 25 meV (time of flight) [Tiedje, 1984] for the conduction band and 48 meV (optical absorption) [Cody et al., 1981] and 51 meV (photoemission yield) [Winer et al., 1988a,b] for the valence band. The model of gap states in a-Si: H proposed by Morigaki et al. [1987] is schematically shown in Fig. 3.2. The important structural defects are the charged Si $T_3^+ + e$ and $T_3^- + h$ centers, the Si-dangling bond (D^0), and the self-trapped hole center (A), in which the superscript of the symbol and its subscript are the charge state and the coordination number of the defects, respectively.

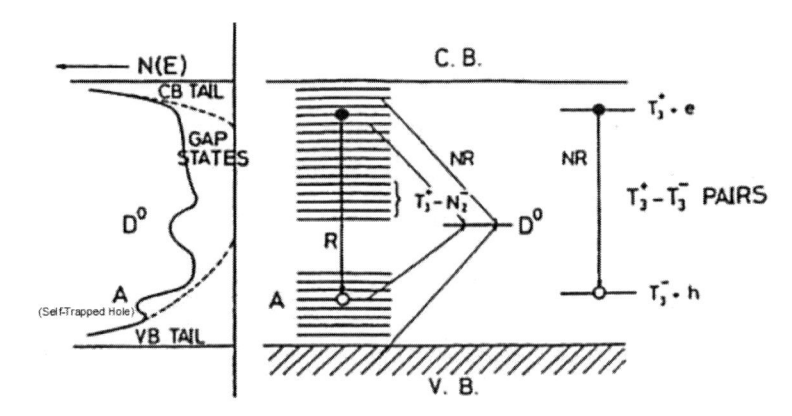

Figure 3.2 Schematic diagram of the tail and gap states and the recombination processes in a-Si:H. R and NR designate radiative and nonradiative recombination, respectively. A diagram of the density of states spectrum is illustrated on the left side. A: self-trapped holes at specific Si–Si weak bonds, D: neutral dangling bonds, T_3^+, T_3^-: positively and negatively charged threefold-coordinated Si atoms, respectively, N_2^-: negatively charged twofold-coordinated nitrogen atom [Morigaki et al., 1987].

The neutral Si-dangling bond in a-Si:H exhibiting an ESR signal with g = 2.0055 has been extensively investigated, particularly, from ENDOR measurements, for example, [Stutzmann and Biegelsen, 1986], [Yokomichi et al., 1987], [Yokomichi and Morigaki, 1993], [Fehr et al., 2012]. The result of ENDOR measurements

[Yokomichi and Morigaki, 1993] will be given later in this section. Those in hydrogenated microcrystalline silicon are also dealt with in Section 4.2.

The light-induced electron spin resonance signals (LESRs) have been observed at $g = 2.004$ and $g = 2.012$ [Street and Biegelsen, 1980]. Street and Biegelsen [1980] proposed that these signals are due to tail electrons and tail holes, respectively. However, with respect to the $g = 2.004$ signal, this has been confirmed to be due to deep centers, because its decay time is over 2 ms, as shown in Fig. 3.3 [Yan *et al.*, 2000]. Note that tail electrons responsible for principal PL have much shorter decay time than it.

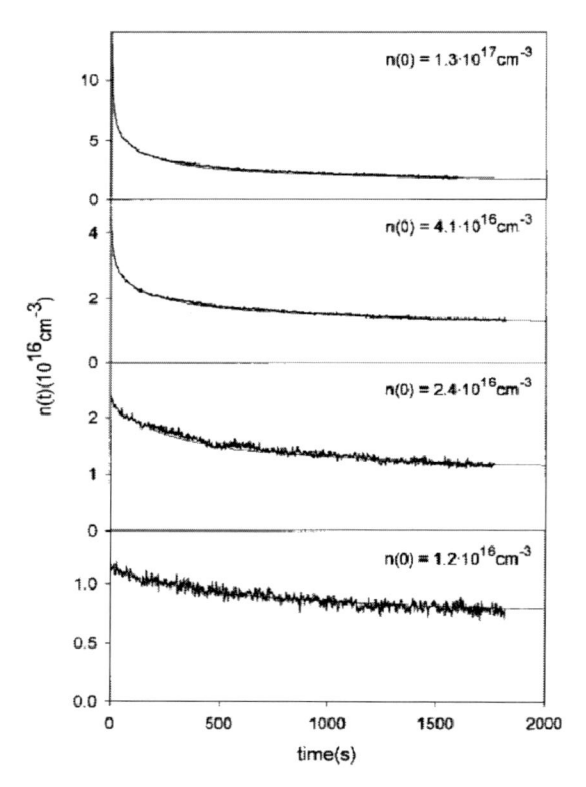

Figure 3.3 Decay of LESR (spin-carrier's density n(t)) in a-Si:H at 40 K after saturation at excitation intensities, from top to bottom, of 19, 2×10^{-1}, 1.6×10^{-3}, and 3.6×10^{-4} mW/cm^2, respectively. Solid lines are theoretical fits to the data. [Reproduced from Yan *et al.*, *Phys. Rev. Lett.*, **84**, 4180 (2000) by permission of American Physical Society.]

Concerning the decay of PL, Aoki *et al.* [2006] have performed its measurements, while concerning the decay of the g = 2.004 signal, Yan *et al.* [2000] have performed its detailed measurements. Thus, now let us consider the long decay component of PL and LESR measured by them [Morigaki, 2009].

Aoki *et al.* [2006] measured spectrally integrated PL intensity decay curves in the region from 10^{-2} to 10^5 s for various generation rates by laser light at 2.37 eV, that is, from 2.0×10^{13} to 2.0×10^{19} cm^3/s, as shown in Fig. 3.4. They fitted these decay curves $I(t)$ *vs.* t with the following function (Williams and Watts law) [Shimakawa, 1985].

$$I_{PL}(t) = I_{PL0}\, t^{\beta-1}\exp[-(t/\tau)^\beta], \tag{3.2}$$

where the estimated values of β and τ are plotted as a function of G in Figs. 3.5(a) and 3.5(b), respectively.

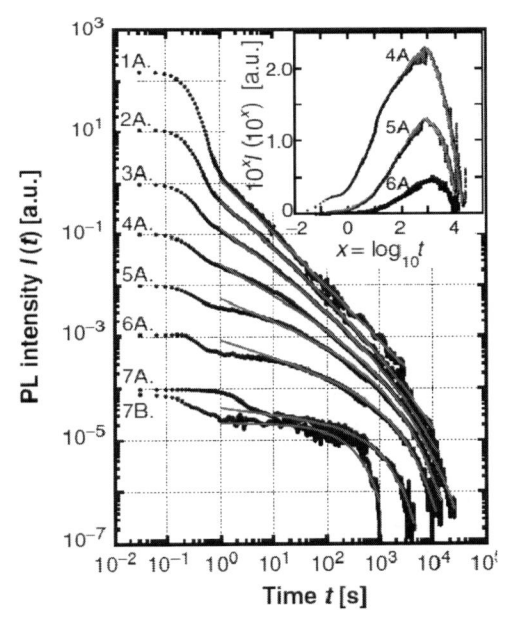

Figure 3.4   Decay of PL intensity in a-Si:H at 3.7 K with various values of G (cm^{3}s^{-1}): 1A: 2.0×10^{19}, 2A: 2.0×10^{18}, 3A: 2.0×10^{17}, 4A: 2.0×10^{16}, 5A: 2.0×10^{15}, 6A: 2.0×10^{14}, 7A: 2.0×10^{13} for film A, 7B: 2.0×10^{18} for film B. The decay curves are fitted to Eq. 3.2. [Reproduced from Aoki *et al.*, *Philos. Mag. Lett.*, **86**, 137 (2006) by permission of Taylor & Francis.]

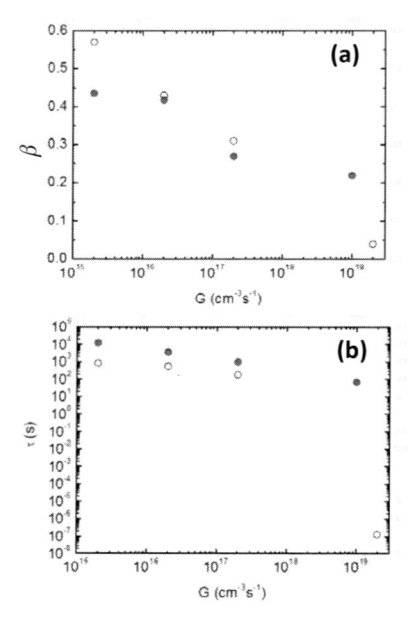

Figure 3.5   (a) β vs. G, (b): τ vs. G. Open circles from PL [Aoki *et al.*, 2006], closed circles from LESR [Yan *et al.*, 2000 ; Morigaki, 2009].

We attempted to fit Yan *et al.*'s decay curves of the LESR intensity with a stretched exponential function, as shown in Fig. 3.6,

$$I_L(t) = I_{L0} \exp[-(t/\tau)^\beta],\tag{3.3}$$

where the estimated values of β and τ are also plotted as a function of G in Figs. 3.5(a) and 3.5(b), respectively. From the above results, it is noted that the values of β and τ for both measurements are different, particularly in τ. In the stretched exponential function, the value of β determines the early decay, while the value of τ governs the decay characteristics in the long-time region. Thus, the small value of β means a rapid decay in the early time. The different values of τ suggest that radiative centers responsible for PL and magnetic centers responsible for LESR are different. According to Aoki *et al.* [2006], the PL component of long lifetime is due to radiative recombination of distant electron–hole pairs in which electrons and holes occupy the tail states of respective bands. In our model, holes are self-trapped holes in weak Si–Si bonds adjacent to Si–H bonds at low temperatures, as will be mentioned in Section 3.2.3.

In our model, magnetic centers responsible for LESR are electrons and holes in deep levels whose detailed models are presented below.

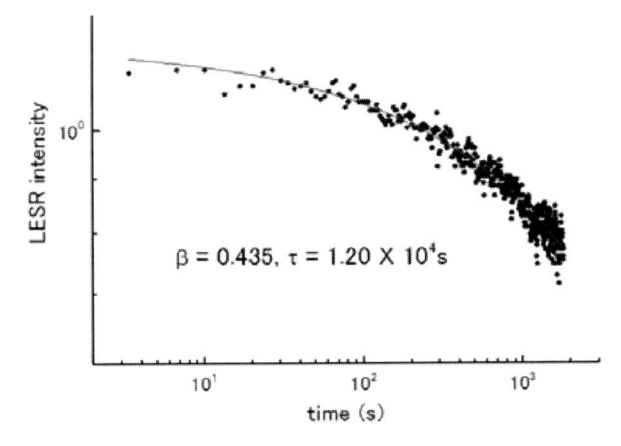

$\beta = 0.435, \tau = 1.20 \times 10^4$s

Figure 3.6   Decay curve of LESR fitted to a stretched exponential function Eq. (3.3) with $\beta = 0.435, \tau = 1.20 \times 10^4$ s for the data corresponding to a lowest curve in Fig. 3.3.

From the above results, we conclude that the magnetic centers responsible for LESR signals are not radiative centers responsible for the long decay PL component, but are nonradiative centers observed as quenching signals at $g_e = 2.004$ and $g_h = 2.013$ in the ODMR measurement [Morigaki, 1984; Morigaki and Yoshida, 1985]. It is noted that shallow tail electrons and self-trapped holes responsible for PL have been identified from time-resolved ODMR [Takenaka *et al.*, 1988] and ODENDOR measurements [Kondo *et al.*, 1989, 1990, 1991; Morigaki and Kondo, 1995], [Morigaki *et al.*, 1995], respectively, as will be mentioned in Section 3.2.3.

The model of the LESR centers proposed by one of the authors [Morigaki, 1983, 2002b] is given as follows: The $g = 2.004$ signal is due to an electron trapped at a T_3^+ center, while the $g = 2.013$ signal is due to a hole trapped at a T_3^- center. These T_3^+ and T_3^- centers are negative-U centers, that is, with negative correlation energy U. It is considered that these centers become stable owing to their environment through strong electron–phonon interaction, because they are deep centers. The creation process of these centers is shown in Fig. 3.7, in which a pair of T_3^+ and T_3^- is created by breaking of a weak Si–Si bond. These pair defects may be separated with each other by thermal energy. These centers have been found to act as nonradiative recombination centers from the ODMR measurement [Morigaki and Yoshida, 1985], in which these centers are observed as quenching signals. It has been found from the ODMR measurement at 2 K that neutral dangling bonds are created under illumination at 2 K, but these charged centers are not created [Han *et al.*, 1987]. From the above model of their creation process, thermal energy is required to separate them with each other, so that the temperature such as 2 K is too low to create their separate pairs. During sample preparation, their separate pairs are created, as enough thermal energy to create them is supplied.

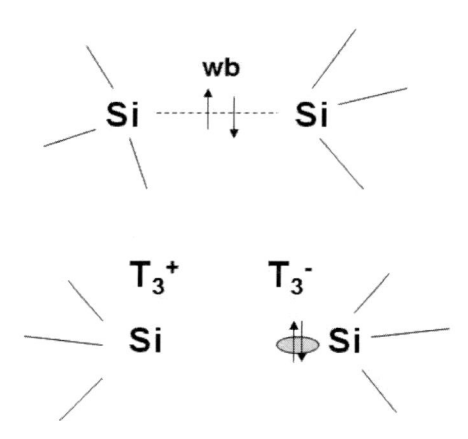

Figure 3.7   Upper part: Before breaking of a weak Si–Si bond. Lower part: After breaking of a weak Si–Si bond. Two centers, T_3^+ and T_3^-, are formed.

LESR measurements [Zhou *et al.*, 1995] indicate that the electron component with $g = 2.004$ at weak optical excitation is slightly increased in intensity with a factor of 1.1 after light soaking at room temperature, while the hole component with $g = 2.013$ is increased with a factor of 2. This result shows unbalance of two signal intensities as a result of light soaking that leads to an increase in the density of neutral dangling bonds. According to LESR measurements by Saleh *et al.* [1992], light-soaked samples exhibit an increase in the hole component compared with as-deposited (or annealed) samples with the same dangling bond density. Thus, it seems to us that an increase in the hole component with a factor of 2 may be mainly due to an increase in the dangling bond density as a result of light soaking. Taking into account this point in the above measurements, these results seem consistent with the creation mechanism of T_3^+ and T_3^- such as mentioned above.

Inequality of the intensities of two components with $g = 2.004$ and $g = 2.013$ has been pointed out and discussed [Schumm *et al.*, 1993; Saleh *et al.*, 1993]. This issue has also been discussed in terms of self-trapping of holes at low temperatures [Morigaki, 1988, 1992].

Umeda *et al.* [1996] suggested from the intensity ratio of two hyperfine components owing to ^{29}Si nucleus in their pulsed ESR measurements on enriched samples that the electron in the antibonding orbital of a weak Si–Si bond extends over two silicon atoms. The deconvolution of the ESR spectra [Morigaki and Hikita, unpublished] into two hyperfine components in its shoulder parts, as shown in Fig. 3.8, led us to the following values of hyperfine interaction components, that is, parallel, $A_{||}$, and perpendicular, $A_\perp$, with ^{29}Si nucleus: $|A_{||}| = 93$ gauss and $|A_\perp| = 51$ gauss. We obtain $|A_{iso}| = 65$ gauss and $|A_{aniso}| = 14$ gauss and then η_0^2 (electron density) $= 0.42$, α_0^2 (s-character percentage) $= 0.10$, and β_0^2 (p-character percentage) $= 0.90$. This result indicates that the magnetic center responsible for the LESR signal with $g = 2.004$ is like a dangling bond center and that its wave function extends over neighboring sites more than that of the dangling bond electron ($\eta_0^2 = 0.54$, $\alpha_0^2 = 0.10$, and $\beta_0^2 = 0.90$ [Yokomichi and Morigaki, 1993]). This result is also consistent with our model that the trapped electron level is shallower than the dangling bond electron level. This is also compared with the result obtained from the electron nuclear double-resonance (ENDOR) measurements [Yokomichi and Morigaki, 1991] on phosphorus-doped a-Si:H, that is, $\eta_0^2 = 0.47$, $\alpha_0^2 = 0.13$, and $\beta_0^2 = 0.87$. This magnetic center is also due to an electron at the negative-U T_3^+ center in our model, although it is generally considered to be due to a conduction-band tail electron. It is interesting to point out that an LESR signal with $g = 1.998$ was observed in μc-Si:H [Kondo *et al.*, 1998] and that it has been identified as due to tail electrons in crystallites in μc-Si:H, as will be mentioned in Section 4.3. A similar signal with $g = 1.987$ has been observed in ODMR measurements, as will be mentioned in Section 3.2.4. From the above consideration, we conclude that the magnetic center responsible for the LESR signal with $g = 2.004$ is a dangling-bond like center whose possible candidate is a negative-U T_3^+ center, as discussed above.

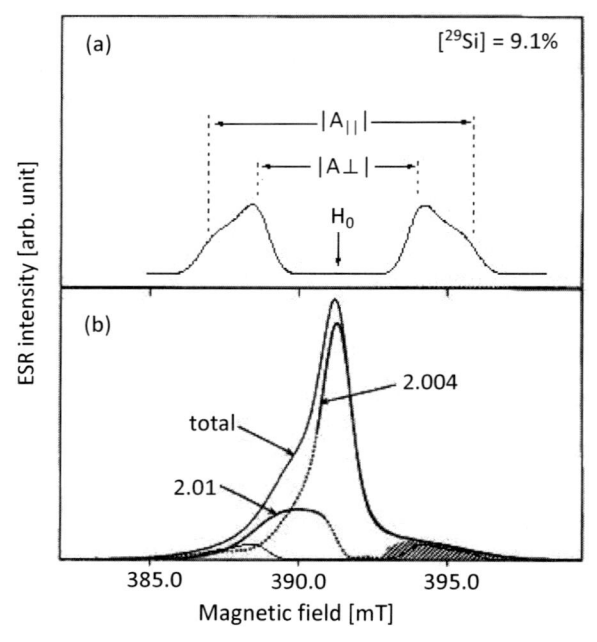

Figure 3.8 Upper part: Simulated hyperfine components (see the text) [Morigaki and Hikita, unpublished, cited in Morigaki, 2002b]. Lower part: Deconvoluted spectra of the LESR of undoped a-Si:H (enriched, [^{29}Si] = 9.1 %) with two components with $g = 2.004$ and 2.01. The shaded region shows the high field hyperfine component. [Reproduced from Umeda *et al.*, *Phys. Rev. Lett.*, **77**, 4600 (1996) by permission of American Physical Society.]

The hole component with $g = 2.013$ has not been investigated in more detail, for example, on hyperfine interaction with ^{29}Si nucleus and so on. Here, we point out only that the values of η_0^2, α_0^2, and β_0^2 for the ESR signal with $g = 2.013$ have been obtained from ENDOR measurements in B-doped a-Si:H as follows: $\eta_0^2 = 0.72$, $\alpha_0^2 = 0.09$, and $\beta_0^2 = 0.91$ [Yokomichi and Morigaki, 1991]. This signal has been suggested to be due to a hole trapped at a negative-U T_3^- center that is also responsible for the hole component of LESR spectra in a-Si:H. From these values, a hole is mainly localized at a central silicon atom and its wave function is like p-function. This is quite reasonable, because the hole is considered to be more deeply trapped than the electron and its character reflects the nature of valence band p-state.

Concerning the conduction-band tail electron and the valence-band tail hole, we here only point out that time-resolved ODMR measurements [Takenaka *et al.*, 1988] detect the tail electron signal with g = 1.987, as mentioned above, and also the self-trapped hole signal with 2.0065 at low temperature, as will be mentioned in Section 3.2.4. Most of tail holes are considered to be self-trapped at weak Si–Si bonds adjacent to a Si–H bond at low temperature, particularly below 60 K [Morigaki, 1984; Morigaki and Kondo, 1995], that is, tail holes form mainly self-trapped holes at low temperature. No tail hole signals have been observed at higher temperature such as at room temperature.

In the following, we compare the g-value of defects among a-Si:H, c-Si, and related materials. Following Watkins and Corbett's model [Watkins and Corbett, 1964], g-shift is described, using the tight-binding approximation for wave function of a dangling bond electron:

$$\Psi = \sum_i \eta_i \Psi_i, \tag{3.4}$$

$$\Psi_i = \alpha_i (\Psi_{3s})_i + \beta_i (\Psi_{3p})_i, \tag{3.5}$$

$$\delta g_{||} = 0, \tag{3.6a}$$

$$\delta g_{\perp} = \lambda \{(1/E_b) - (1/E_a)\} \eta_0^2 \beta_0^2, \tag{3.6b}$$

where η_i, α_i^2, β_i^2, E_a, and E_b are the electron density, s-character percentage, p-character percentage at ith site, the energy separation between the antibonding orbital level and the dangling bond level and between the dangling bond level and the bonding orbital level, respectively, and the direction of $||$ is defined by that of the dangling bond. The 0th site is a dangling bond site. As seen from Eq. (3.6b), $\delta g_{\perp}$ is determined by the p-character percentage, β_0^2, of the wave function and the electron density, η_0^2, at the dangling bond site except by E_a and E_b.

In the following, we compare the observed g-shifts, that is, $\Delta g_{||}$ and $\delta g_{\perp}$, with the theoretical prediction shown in Eqs. (3.6a) and (3.6b). A diagram of $\delta g_{||}$ vs. $\delta g_{\perp}$ is shown in Fig. 3.9, in which the observed values of normal dangling bonds and hydrogen-related dangling bonds in a-Si:H [Morigaki *et al.*, 1998; Morigaki and

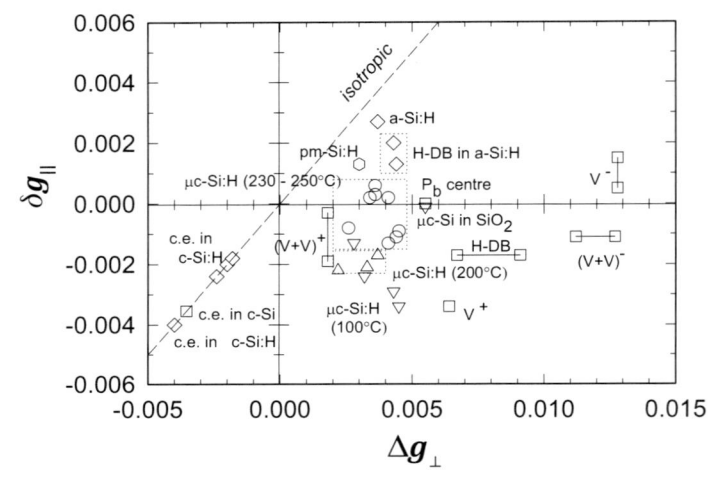

Figure 3.9 A diagram of $\delta g_{||}$ vs. $\delta g_{\perp}$. Square: defects in c-Si; V^+, V^-, $(V + V)^+$, $(V + V)^-$, and H–DB designate positively charged and negatively charged single vacancies [Watkins, 2000], positively charged and negatively charged divacancies [Watkins, 2000], and vacancy–hydrogen complex [Nielsen *et al.*, 1997]. Circle: anisotropic centers in HWCVD μc-Si:H prepared at 230–250°C [Morigaki *et al.*, 2002]. Triangle: anisotropic centers in HWCVD μc-Si:H prepared at 200°C [Morigaki *et al.*, 2002]. Diamond (a-Si:H): normal dangling bonds [Morigaki and Hikita, 1999]. Diamond (H–DB in a-Si:H): hydrogen-related dangling bonds [Morigaki and Hikita, 1999; Morigaki *et al.*, 2003]. Hexagon (pm-Si:H): anisotropic centers [Morigaki *et al.*, 2002]. Inverse triangle (μc-Si:H): anisotropic centers (PECVD at 100°C) [Morigaki *et al.*, 2003]. Square: P_b centers [Nishi, 1971] and anisotropic centers in μc-Si:H [Kondo *et al.*, 2000]. Inverse triangle (μc-Si in SiO_2) : anisotropic centers [Ehara *et al.*, 2000]. For anisotropic centers of orthorhombic symmetry, two components of the g-tensor with similar values are shown by two points connected with a line. For conduction electron in μc-Si:H, three points (square) are shown [Morigaki *et al.*, 2003]. Diamond: conduction electron in c-Si [Wilson and Feher, 1961].

Hikita, 1999] are plotted as well as those of defects for hydrogenated microcrystalline silicon (μc-Si:H) [Morigaki *et al.*, 2002], hydrogenated polymorphous silicon (pm-Si:H) [Morigaki *et al.*, 2002], and c-Si [Watkins, 1965; Nielsen *et al.*, 1997]. For comparison, $\delta g_{||}$ and $\delta g_\perp$ of intrinsic defects created by radiation damage and also by proton implantation [Nielsen *et al.*, 1997] are shown as well as those of conduction electrons in μc-Si:H [Finger *et al.*, 1994] and c-Si [Wilson and Feher, 1961] and the P_b center in Si–SiO$_2$ interface [Nishi, 1971]. The magnetic centers in μc-Si: H [Morigaki *et al.*, 2002], [Kondo *et al.*, 2000], [Morigaki *et al.*, 2003, 2009b] and μc-Si in SiO$_2$ [Ehara *et al.*, 2000] are anisotropic centers involving a dangling bond located on the boundary between microcrystalline phase and amorphous phase. The wave function of a dangling bond of the P_b center is localized on the central silicon atom with $\eta_0^2 = 0.8$, $\alpha_0^2 = 0.12$, and $\beta_0^2 = 0.88$ [Brower, 1986]. In c-Si, g-shifts of defects have generally an orthorhombic symmetry, so that the largest values among δg components, δg_x, δg_y, and δg_z, are plotted on the $\delta g_\perp$ axis and others on the $\delta g_{||}$ axis.

As seen in Fig. 3.9, the experimental points for defects in c-Si are almost either along or near the $\delta g_\perp$ axis except for positively charged single vacancy V$^+$. In Fig. 3.9, $\delta g_{||}$ and $\delta g_\perp$ of LESR centers are not plotted, because these have not been estimated, although only isotropic g-shifts have been obtained, assuming that the g-values are isotropic. However, some points including μc-Si:H deviate from $\delta g_{||} = 0$ to $\delta g_{||} < 0$. This is qualitatively accounted for as follows: Eqs. (3.6a) and (3.6b) are obtained within the extent of the wave function only over four nearest neighbors. According to Watkins and Corbett [1964], the interatomic currents arising from the overlapping of the wave function on further neighboring sites contribute somewhat to $\delta g_{||}$ and $\delta g_\perp$, but the sign of $\delta g_{||}$, that is, deviation of $\delta g_{||}$ from the free electron value of 2.0023, is generally unpredictable. Thus, the result of $\delta g_{||} < 0$ means that the wave function extends over further neighboring sites beyond four nearest neighbors. In Fig. 3.9, most of μc-Si:H were prepared by hot-wire chemical vapor deposition (HWCVD) at either 200°C or 230–250°C [Morigaki *et al.*, 2002]. One point was obtained for a sample prepared by plasma-enhanced CVD (PECVD) at 100°C [Morigaki *et al.*, 2003, 2009b], using mixture gas of silane and hydrogen with mixture gas ratio γ, that is, (SiH$_4$)/(SiH$_4$) + (H$_2$)] =

2.2%, whose $\delta g_{\parallel}$ deviates very much from the $\delta g_{\parallel} = 0$ axis. This deviation indicates that the wave function of an unpaired electron extends over further neighboring sites, as mentioned above. We note that the experimental point of $\delta g_{\parallel}$ and $\delta g_{\perp}$ for PECVD μc-Si:H prepared at 250°C with $\gamma = 2.0\%$ [Kondo *et al.*, 2000] almost coincides with that for the Pb center. Thus, this indicates that $\delta g_{\parallel}$ and $\delta g_{\perp}$, particularly $\delta g_{\parallel}$, for this sample are different from those for PECVD μc-Si:H prepared at 100°C with $\gamma = 2.2\%$, namely they are dependent on deposition temperature.

As seen in Fig. 3.9, it is interesting that the experimental point for the hydrogen-related dangling bond (H–DB in a-Si:H) differs from the point for the vacancy–hydrogen complex (H–DB). This suggests that environment around these centers is different between two centers. In a-Si:H, the local network around the dangling bond can be relaxed, whereas in c-Si, the local structure around the dangling bond is almost rigid, as the distance between hydrogen and dangling bond site is almost the same as that calculated [Nielsen *et al.*, 1997], assuming unrelaxed geometry of the vacancy–hydrogen complex, as was mentioned before.

Further we point out that $\delta g_{\parallel}$ and $\delta g_{\perp}$ of the dangling bond (normal dangling bond) in a-Si:H are different from those of the P_b center and the defects in c-Si, as shown in Fig. 3.9. The wave function of the dangling bond electron in a-Si:H is localized on the central silicon atom with $\eta_0^2 = 0.54$ [Yokomichi and Morigaki, 1993]. This means that it extends over neighboring sites more than the P_b center ($\eta_0^2 = 0.80$) [Brower, 1986]. Thus, it is suggested that difference in η_0^2 between the normal dangling bond center in a-Si:H and the P_b center results in $\delta g_{\parallel}$ more than 0 for the former center, taking into account that the percentages of s-character and p-character in both centers are similar to each other. This is also related to local relaxation around the dangling bond. The above consideration is very qualitative, but a quantitative consideration is required.

As mentioned in Section 1.4, the ODMR measurement provides a powerful tool for a study of tail states and structural defects in a-Si:H [Morigaki, 1983, 1984; Cavenett *et al.*, 1983; Morigaki and Kondo, 1995]. Some of ODMR and ODENDOR measurements are presented in the next section.

3.2.3 Self-trapping of Holes

In this section, we consider the self-trapping of holes in a-Si: H. As mentioned in Section 3.2.2, a hole may be self-trapped at a weak Si–Si bond adjacent to a Si–H bond. Such so-called self-trapping of holes is a kind of extrinsic self-trapping having been extensively considered by Toyozawa [1986] and Shinozuka and Toyozawa [1979]. We have already pointed out the reasons of why a hole is self-trapped at a weak Si–Si bond adjacent to a Si–H bond in a-Si:H as follows [Hirabayashi *et al.*, 1980]: The Si–H bond has a stronger bonding energy than that of Si–Si bond and the electronegativity of hydrogen is larger than that of silicon, so that the bonding electrons of silicon may be polarized toward hydrogen and, as a result, one of four Si–Si bonds adjacent to a Si–H bond becomes weak. This has also been suggested from *ab initio* molecular dynamics computer simulation by Yonezawa *et al.* [1991].

The existence of self-trapped holes at low temperatures has been suggested from photomodulation-infrared (IR) spectra in a-Si: H. The spectra have been observed as a change in the prove-light transmittance under optical excitation, that is, pumping light. In a-Si:H, it has been found by Vardeny and Olszakier [1987] that the IR peak intensity at 2000 cm^{-1} decreases under optical excitation. This result has been accounted for by considering that the dipole moment for this IR mode becomes small by self-trapping of a hole in a weak Si–Si bond adjacent to the Si–H bond responsible for the IR mode. This is in contrast with the change in the IR peak intensity at 2000 cm^{-1} by prolonged illumination that it increases after illumination [Kong *et al.*, 1995; Zhao *et al.*, 1995]. The former corresponds to the IR mode under illumination, but the latter corresponds to that after illumination. Under illumination, self-trapped holes in the weak Si–Si bonds affect the effective charge of the stretching vibration of their nearby Si–H bonds to be reduced compared with without illumination. After illumination, this effect does not occur, although the network receives a structural change [Fritzsche, 1995; Nonomura *et al.*, 2000; Morigaki, 2001].

Very recently, Oheda [1999] has found that the photomodulated IR peak intensity around 2000 cm^{-1} is affected by prolonged illumination at 297 and 13 K, namely the amount of the decrease in the IR peak intensity decreases after prolonged illumination by a Kr$^+$ laser light of 1.92 eV, as shown in Fig. 3.10. The decrease is more enhanced at the low energy side around 1950 cm^{-1} of the 2000 cm^{-1} peak than at the higher energy side. The low energy side around 1950 cm^{-1} seems to correspond to the 1935 cm^{-1} band observed by Zhao *et al.* [1995]. The 1935 cm^{-1} band is assumed to be due to a more elongated weak Si–Si bond adjacent to a Si–H bond [Morigaki, 2001], that is, it is assumed that there are two types of weak Si–Si bonds, that is, normal weak bonds and more elongated weak bonds. We note that the photomodulation-IR absorption spectra at 2000 cm^{-1} may be due to the stretching vibration mode of a Si–H bond adjacent to a normal weak Si–Si bond. Thus, the distance between hydrogen and silicon being located at the further end of the weak Si–Si bond from hydrogen is longer than a normal weak Si–Si bond adjacent to a Si–H bond,

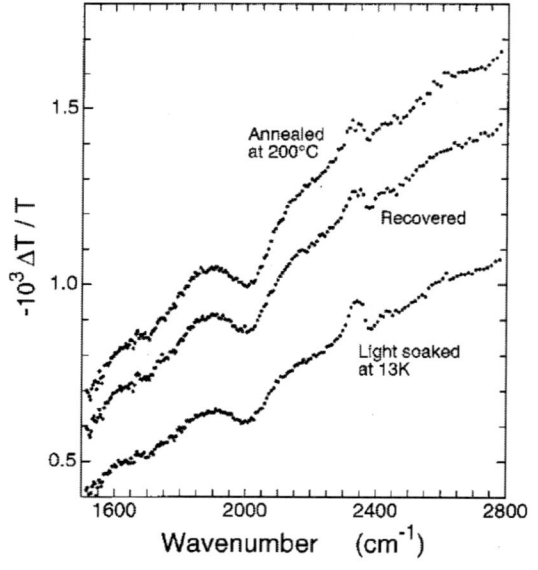

Figure 3.10 Photomodulation spectra of a-Si:H in the annealed and light-soaked states (at 13 K for 20 h by a Kr$^+$ laser light at 1.9 eV and 2 W cm^{-2}). The annealed state was measured after keeping the sample at room temperature for 2 days after the light soaking at 13 K. [Reproduced from Oheda, *Phys. Rev. B*, **60**, 16531 (1999) by permission of American Physical Society.]

so that the screening effect of the self-trapped hole on the Si–H bond becomes weaker than the case of the normal weak bond. Here, we should say that the self-trapped hole is more localized at a Si site of the further end of the weak Si–Si bond from hydrogen than at a Si site of the Si–H bond [Morigaki *et al.*, 1995]. This means that the decrease in the intensity of the photomodulation-IR absorption spectrum at 1950 cm^{-1} by optical excitation is smaller than the case of the 2000 cm^{-1} band. This is the case in Oheda's measurements, that is, more elongated weak Si–Si bonds mentioned above are created by prolonged illumination, so that a different photomodulation-IR spectrum from that before illumination is observed, emphasizing the part around 1950 cm^{-1}, as mentioned above.

The hole's self-trapping model can account for the temperature dependence of drift mobility of holes as follows: The drift mobility of holes has been observed above 200 K and it decreases with the hydrogen content exponentially [Das *et al.*, 1991]. This can be understood by taking into account that holes are more and more self-trapped with decreasing temperature below 200 K, so that their drift mobility is difficult to be observed. Furthermore, the density of weak Si–Si bonds adjacent to Si–H bonds increases with the hydrogen content.

The microscopic structure of self-trapped holes has been investigated by means of optically detected ENDOR measurement [Kondo *et al.*, 1990, 1991, 1993; Morigaki *et al.*, 1995]. The hole is assumed to be centered at site b extending around site b in Fig. 3.11, because the electron in the weak Si–Si bond tends to extend toward hydrogen site owing to a difference in electronegativity between Si and H. As a result, the hole is displaced toward site b.

The wavefunction of the hole is expressed in the tight-binding approximation, as Eqs. (3.4) and (3.5). The ODENDOR spectra are shown in Figs. 3.12(a) and 3.12(b). The solid line is the calculated line adjusted to be fitted to the experimental curve with the values of the following parameters [Morigaki *et al.*, 1995]: $\eta_0^2 = 0.13$, $\alpha_0^2 = 0.03$, and $\beta_0^2 = 0.97$, where the suffix 0 designates the site b. From the result, the hole wavefunction has almost p-character and the relative hole density at the site b is 13%.

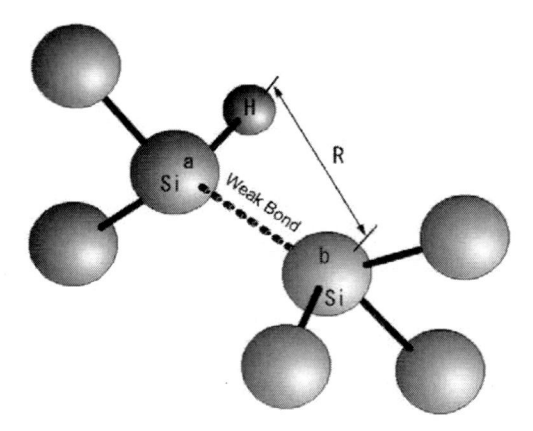

Figure 3.11 Atomic configuration around a specific weak Si–Si bond consisting of a weak Si–Si bond and Si–H bond.

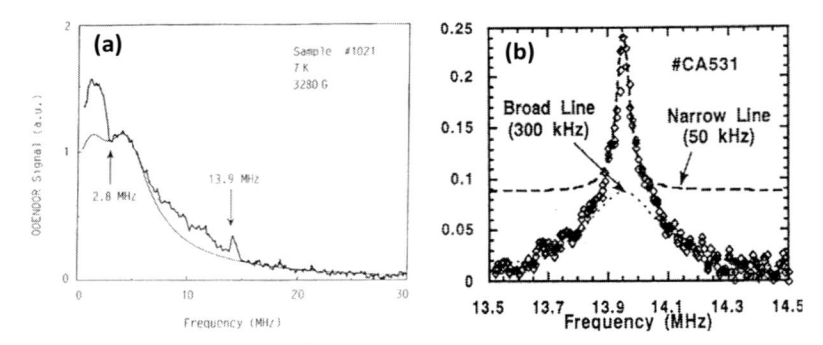

Figure 3.12 (a) ODENDOR spectrum of a-Si:H sample prepared at 250°C. Its hydrogen content is about 15 at.%. The solid curve is a calculated spectrum (see the text). The positions of natural nuclear frequencies of ^{29}Si and ^{1}H are shown at 2.8 and 13.9 MHz, respectively. [Reproduced from Morigaki *et al.*, *J. Non-Cryst. Solids*, **190**, 38 (1995) by permission of Elsevier]. (b) ODENDOR spectrum of a-Si:H sample prepared at 300°C. Its hydrogen content is about 3%. The dotted and dashed curves correspond to the broad line and narrow line, respectively. The FWHM of each line is 300 kHz and 50 kHz, respectively. [Reproduced from Kondo and Morigaki, *J. Non-Cryst. Solids*, **137 & 138**, 247 (1991) by permission of Elsevier.]

The ODENDOR signal at 13.9 MHz shown in Figs. 3.12(a) and 3.12(b) corresponds to the natural frequency of hydrogen nucleus, but its linewidth is wide ($\approx$290 kHz) compared with the normal

NMR linewidth (≈several kHz). This ENDOR signal is called the matrix ENDOR signal. The linewidth is given by dipolar interaction between the hole spin and hydrogen nuclear spin and it is estimated from the linewidth that the distance between hydrogen and b site is 4 Å. This means that the weak Si–Si bond is elongated with 25% for the normal distance 2.35 Å. This clearly indicates that as a hole is self-trapped at the weak Si–Si bond, the bond length is elongated from the normal bond length. Under illumination, Yonezawa *et al.'s ab initio* molecular dynamics calculation [Yonezawa *et al.*, 1991] and Jones and Lister's calculation [Jones and Lister, 1990] based on the local functional method gave 10% and 40%, respectively. These are reasonably compared with the experimental estimate (25%).

The energy level of self-trapped holes in a-Si:H has been estimated from PA measurements. The level of self-trapped holes is located at 0.25 eV above the edge of the valence band for high-quality a-Si:H samples and at 0.35 eV for low-quality a-Si:H samples containing a large amount of hydrogen ([H] ≈ 30 at.%) [Hirabayashi and Morigaki, 1983b; Yamaguchi and Morigaki, 1991; Morigaki *et al.*, 1995].

3.2.4 Tail Electron States

Tail electron states have been investigated by time-resolved ODMR measurements in a-Si:H. As will be mentioned in Section 3.3, the tail electron has a lifetime of ~1 μs, depending upon its distance from tail holes (or self-trapped holes at low temperatures). Thus, we attempted to observe the ESR signal of tail electrons by using time-resolved technique of ODMR, that is, a pulsed optical excitation, a pulsed microwave, and a dual-channel boxcar integrator at 2 K [Takenaka *et al.*, 1988]. As optical excitation, an N_2 pumped dye laser of 5 ns width and a peak power of 220 kW/cm^2 at 510 nm were used. A pulsed microwave of 1–10 μs was applied to samples and a detection gate for observing ODMR signals was delayed from the pulsed laser light and the pulsed microwave. The timing of the laser pulse, microwave pulse, and gate pulse is illustrated in Fig. 3.13. The ODMR signal can be observed as the output signal difference between gates A and B. The observed ODMR signal monitoring the total emitted light is shown in Fig. 3.14 for various delay time, t_d (see Fig. 3.13). Two

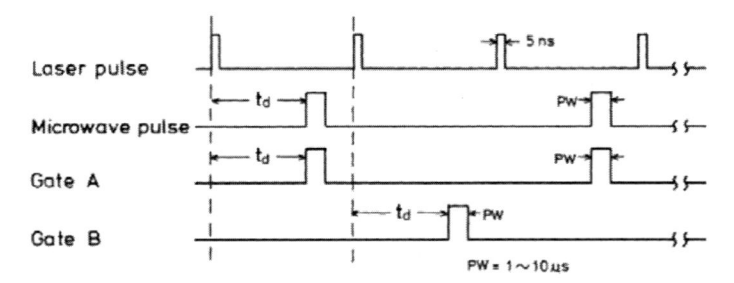

Figure 3.13 Timing of the laser pulse, microwave pulse, and gate pulse for observing the enhancing signal. [Reproduced from Takenaka *et al.*, *J. Phys. Soc. Jpn.*, **57**, 3858 (1988) by permission of The Physical Society of Japan.]

components are seen at t_d = 50 μs in Fig. 3.14, in which one of them with g = 2.0041 is the main signal corresponding to that due to self-trapped holes, as mentioned in Section 3.2.3, while the other with g = 1.989 (the shoulder signal) is due to tail electrons. As seen from Fig. 3.14, the ODMR signal due to tail electrons decays more quickly than that due to self-trapped holes. The former signal is observed only for delay time shorter than 100 μs. The g-value and the linewidth (FWHM) are shown as functions of delay time in

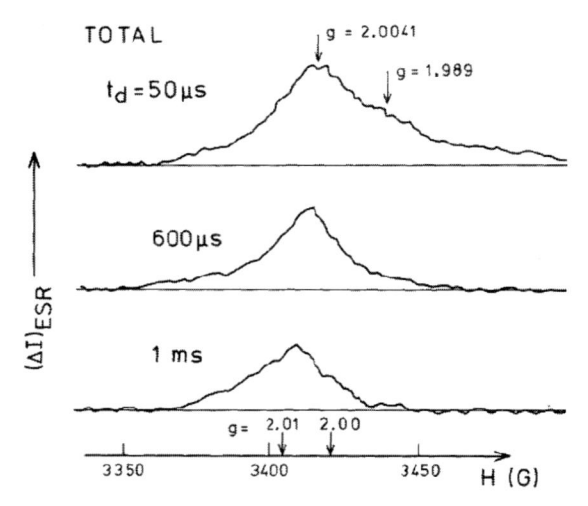

Figure 3.14 TRODMR spectra of a-Si:H for various delay times. Total emitted light was monitored. [Reproduced from Takenaka *et al.*, *J. Phys. Soc. Jpn.*, **57**, 3858 (1988) by permission of The Physical Society of Japan.]

Figs. 3.15 and 3.16, respectively, in which the g-values corresponding to the peak positions of the main signal and the shoulder signal and their linewidth obtained by monitoring monochromatized emission light at 1.28 eV are also shown and g_A designates the g-value of the main signal due to self-trapped holes. As seen in Fig. 3.15, the g-values of the main signal become smaller as delay time becomes shorter. This result can be accounted for in terms of exchange interaction between the tail electron and the self-trapped hole. The average g-value of these two centers, that is, $(g_A + g_e)/2$ (g_e is the g-value of tail electrons), can be seen for short delay time owing to the exchange interaction, as shown in Fig. 3.15. For long delay time, two signals are observed at g-values of each isolated center. From Fig. 3.15, the g-value of the isolated tail electron is estimated to be 1.987. The exchange interaction affects the linewidth *vs.* delay time, as shown in Fig. 3.16. The result of the main signal as a function of delay time (exchange broadening)

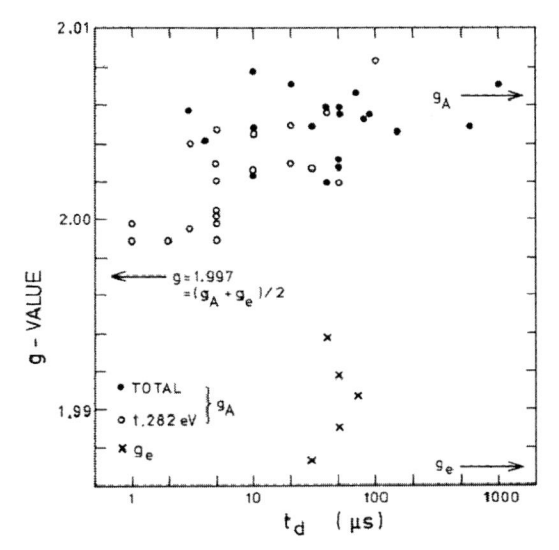

Figure 3.15 Dependence of g-values of the self-trapped holes, g_A, at specific weak Si–Si bonds and the tail electrons, g_e, on t_d in a-Si:H. The solid circles and crosses designate the experimental points obtained by monitoring total emitted light, while the open circles show those obtained by monitoring monochromatized emitted light at 1.282 eV. [Reproduced from Takenaka *et al., J. Phys. Soc. Jpn.*, **57**, 3858 (1988) by permission of The Physical Society of Japan.]

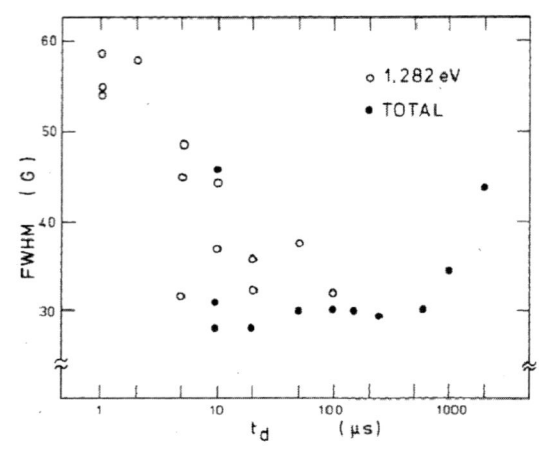

Figure 3.16 Dependence of FWHM (full width at half-magnitude) of the enhancing signal due to self-trapped holes on t_d in a-Si:H. [Reproduced from Takenaka *et al., J. Phys. Soc. Jpn.*, **57**, 3858 (1988) by permission of The Physical Society of Japan.]

is consistent with this picture. The results of the time-resolved ODMR measurement are also consistent with the model of principal PL that arises from radiative recombination between the tail electron and the self-trapped hole at low temperatures (see Section 3.3).

3.2.5 Hydrogen-related Dangling Bonds

In Section 3.5.4, the existence of two types of Si-dangling bonds is shown along with the experimental evidence. The hydrogen-related dangling bond is a dangling bond having a nearby hydrogen, as shown in Fig. 3.17. It has been observed in ENDOR measurements [Yokomichi and Morigaki, 1987, 1993], as the observation of a local ENDOR signal, as shown in Section 3.5.4. At the same time, it is interesting to point out the results of deconvolution of the ESR line due to Si-dangling bonds. The ESR spectrum of hydrogen-related dangling bonds exhibits two hyperfine components due to nuclear spin (magnitude = 1/2) of ^{1}H. A schematic hyperfine pattern due to ^{1}H is shown in Fig. 3.18(b), where $A_{\parallel}$ and $A_{\perp}$ are hyperfine interaction constants parallel and perpendicular to the principal axis of the axially symmetric hyperfine interaction tensor, respectively. Actually, the spin packet component is broadened,

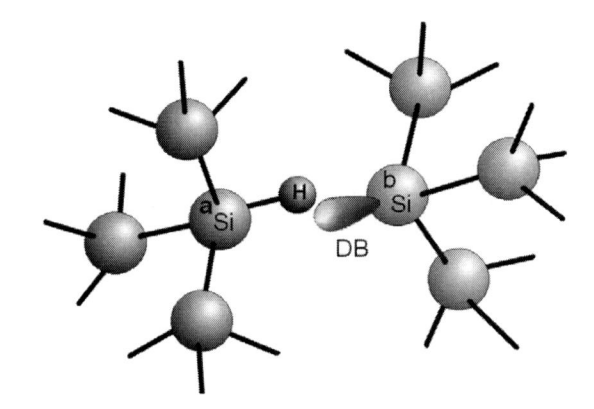

Figure 3.17 Schematic view of the hydrogen-related dangling bond.

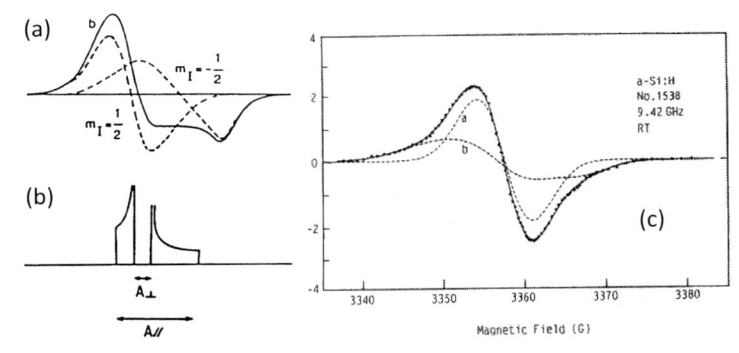

Figure 3.18 (a) Hyperfine structure (hfs) of the hydrogen-related dangling bond due to the ^1H nucleus. Two components, $m_I = 1/2$ and $-1/2$, are shown. [Reproduced from Hikita *et al.*, *J. Phys. Soc. Jpn.*, **66**, 1730 (1997) by permission of The Physical Society of Japan.] (b) Schematic diagram of hfs with axial symmetry due to the ^1H nucleus. [Reproduced from Hikita *et al.*, *J. Phys. Soc. Jpn*, **66**, 1730 (1997) by permission of The Physical Society of Japan.] (c) Deconvolution of the observed ESR spectrum for a-Si:H into two components due to the normal dangling bond a and the hydrogen-related dangling bond b. [Reproduced from Morigaki *et al.*, *J. Non-Cryst. Solids*, **227–230**, 338 (1998) by permission of Elsevier.]

so that an observed spectrum is expected to be broadened like Fig. 3.18(a). We have attempted several samples prepared at different gas ratio of SiH_4 and H_2 at 100°C by PECVD [Morigaki *et al.*, 1998]. An example of deconvolution is shown in Fig. 3.18(c). From the deconvolution, the values of isotropic hyperfine interaction constant with hydrogen nucleus A_{iso} and anisotropic hyperfine interaction constant with hydrogen nucleus A_{aniso} are estimated [Hikita *et al.*, 1997; Morigaki *et al.*, 1998]. The isotropic hyperfine interaction constant arises from the Fermi-type contact interaction, while the anisotropic hyperfine interaction constant arises from dipolar interaction between the hole spin and the hydrogen nuclear spin. Thus, these constants are described as follows:

$$A_{iso} = A_1 \exp(-2R/a^*), \tag{3.7}$$

$$A_{aniso} = A_2/R^3 a^*, \tag{3.8}$$

where R is the distance between hydrogen site and site b in Fig. 3.17. The experimental result is plotted as A_{iso} vs. A_{aniso} in Fig. 3.19. The experimental points are well fitted by a curve obtained from Eqs. (3.7) and (3.8), as shown below:

$$A_{iso} = \exp(\alpha - \beta A_{aniso}^{-1/3}), \tag{3.9}$$

$$\alpha = \ln A_1, \tag{3.10}$$

$$\beta = 2A_2^{1/3}/a^* \tag{3.11}$$

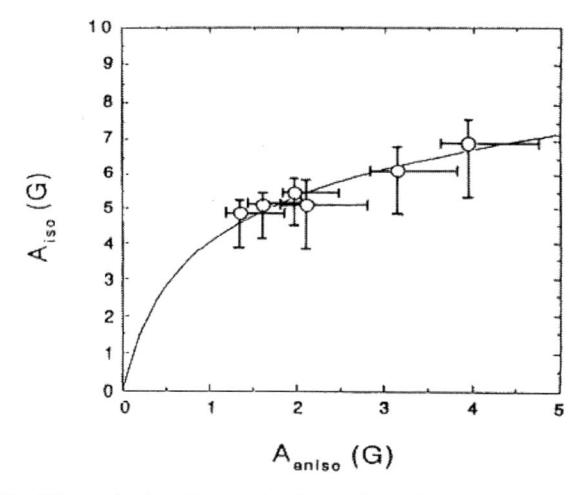

Figure 3.19 Plot of the isotropic hyperfine interaction constant *vs.* anisotropic hyperfine interaction constant in a-Si:H. The solid line is a calculated one following to Eq. 3.9. [Reproduced from Morigaki *et al.*, *J. Non-Cryst. Solids*, **227–230**, 338 (1998) by permission of Elsevier.]

The fitted values of α and β to the A_{iso} vs. A_{aniso} curve shown in Fig. 3.19 are 2.80 G and 1.41 $G^{1/3}$, respectively. Using a theoretical value of $A_2 = 3.781 \times 10^{-23}$ G cm^3, we obtain $a^* = 4.7$ Å. The wavefunction of the dangling bond electron is normally given by a linear combination of atomic 3s and 3p orbitals of Si. Thus, Eq. (3.7) is an approximate expression of the isotropic hyperfine interaction constant. This means that a^* is an approximate extent of the wavefunction of the dangling bond electron.

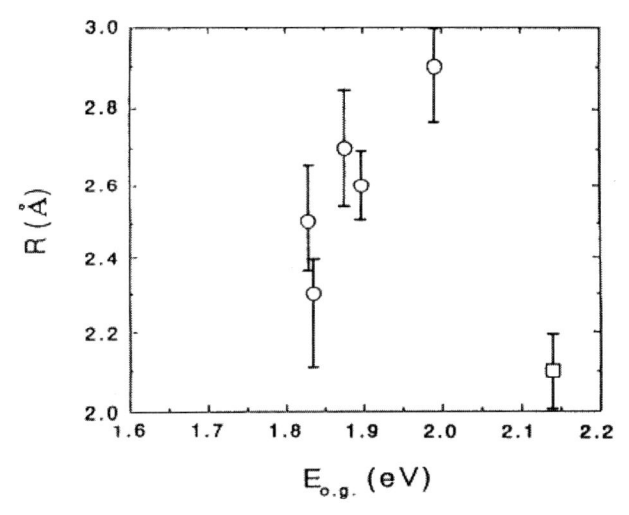

Figure 3.20 Distance between a dangling bond site and hydrogen in the hydrogen-related dangling bond, R, as a function of optical gap energy, E_{og}, in a-Si:H. An open square is the experimental point for sample No. 1527 containing a microcrystalline phase (crystalline volume fraction = 0.51). [Reproduced from Morigaki *et al.*, *J. Non-Cryst. Solids*, **227–230**, 338 (1998) by permission of Elsevier.]

The values of R estimated from A_{aniso} range between 2.1 Å and 2.9 Å, depending on γ. R is plotted as a function of optical gap energy, as shown in Fig. 3.20. The percentage densities of normal dangling bonds N_a and hydrogen-related dangling bonds N_b relative to the total density are shown as functions of gas-dilution ratio γ in Fig. 3.21. As seen in Fig. 3.20, R decreases with decreasing E_{og} except for the sample containing the microcrystalline phase (symbol square in Fig. 3.20). R seems to be correlated with the hydrogen content in such a way that, as the hydrogen content decreases, R decreases. When the sample contains a large amount of hydrogen, local relaxation around the dangling bond site easily occurs owing to flexibility of the amorphous network. This leads to elongation of the distance between the dangling bond site and hydrogen. For the sample containing the microcrystalline phase (crystalline volume fraction X_c = 0.51), R is 2.12 Å. This sample is expected to contain a large amount of hydrogen from its optical gap energy. However, this result is in contrast with other samples, as seen in Fig. 3.20. This difference may be related to the presence of the microcrystalline phase.

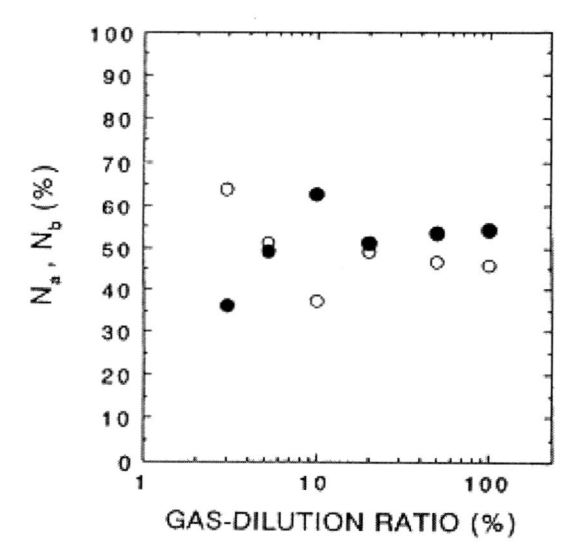

Figure 3.21 Percentage densities of normal dangling bonds (open circles) and hydrogen-related dangling bonds (closed circles) as a function of gas-dilution ratio (%) in a-Si:H. [Reproduced from Morigaki *et al.*, *J. Non-Cryst. Solids*, **227–230**, 338 (1998) by permission of Elsevier.]

In the following, we discuss the percentage densities of N_a and N_b. In the range 10% $\leq \gamma \leq$ 100%, N_b is greater than N_a, consistent with the amount of hydrogen contained in samples. An increase in N_b at γ = 100% seems to correspond to an increase in hydrogen content. Below γ = 5%, N_a becomes greater than N_b. This decrease correlates with the presence of the microcrystalline phase, although the hydrogen content seems to increase with decreasing

γ as expected from the optical gap energy. This effect is also consistent with the value of R for sample with $X_c = 0.51$, as mentioned above.

We note that the samples discussed here were prepared at 100°C, so that they are not high-quality samples and contain a large amount of hydrogen. In high-quality samples, N_a seems to be much greater than N_b, that is, dangling bonds are normally normal dangling bonds (see Section 3.5.4).

It is interesting to note that vacancy–hydrogen complexes shown in Fig. 3.22 have been observed in c-Si implanted with hydrogen at low temperatures (≤ 100 K) [Bech Nielsen *et al.*, 1997]. Hyperfine structures due to a nearby hydrogen nucleus have been observed, in which the distance between dangling bond site and hydrogen is obtained to be 2.7 Å. In the diamond structure, the distance is estimated to be 2.8 Å, assuming 1.5 Å for the Si–H bond length. This is quite close to the observed distance in a-Si:H. A detailed description on the vacancy–hydrogen complex will be given in Section 3.5.4.

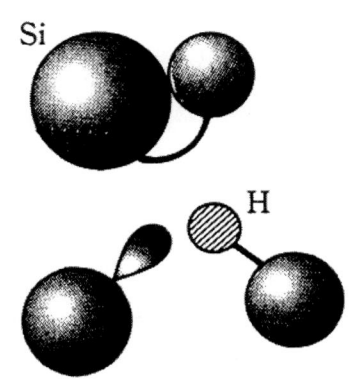

Figure 3.22 Atomic configuration of the vacancy–hydrogen complex in c-Si. [Model adapted from Nielson *et al.*, 1997.]

3.3 Recombination Processes

3.3.1 *Introduction*

In the case of a-Si:H, the electrons excited to the conduction band and holes created simultaneously in the valence band are thermalized to the tail states of the respective bands. Some carriers are trapped at deep gap states. Self-trapping of holes at low temperatures has also been suggested, as mentioned in Section 3.2.3. The recombination of carriers in a-Si:H is completed by radiative or nonradiative transitions, which is much slower than the thermalization. Therefore, various localized states in the band gap, for example, tail states, gap states created by defects, and self-trapped holes, are involved in the recombination processes. Thus, the study of the recombination processes is important to understand the nature of the localized states in the band gap. Detailed information of the recombination processes in a-Si:H has been obtained by measurements of PL and those combined with magnetic resonance, that is, ODMR measurements. See also [Morigaki, 2002a].

3.3.2 Photoluminescence in a-Si:H

In 1973, Engemann and Fischer [1973] observed strong PL in a-Si:H for the first time. The quantum efficiency estimated for the PL at low temperature in a high-quality a-Si:H was of the order of unity [Engemann and Fischer, 1974]. Such strong PL is not observed in c-Si, which is known to be an indirect gap semiconductor. The radiative transition from the bottom of the conduction band to the top of the valence band is forbidden by the k-selection rule in indirect gap semiconductors. As described above, the recombination of carriers occurs by transitions between the localized states. In this case, the transitions are not affected by the k-selection rule. This fact results in the strong PL in a-Si:H. Examples of the PL spectra in a-Si:H films are shown in Fig. 3.23. In the case of a-Si:H, the transitions between the conduction and valence bands are not forbidden by the k-selection rule, because the k-selection rule is relaxed by lack of long-range order in the amorphous network. However, the hot luminescence of the energy higher than band gap (1.7–1.8 eV) is not observed in a-Si: H, as the carriers are thermalized much more quickly than the radiative transition. Normally, the cooling rate is expected to be approximately $\hbar\omega^2$ where ω denotes the frequency of the phonons. However, recent measurements have shown a faster cooling rate of 2 eV/ps [Wraback and Tauc, 1992].

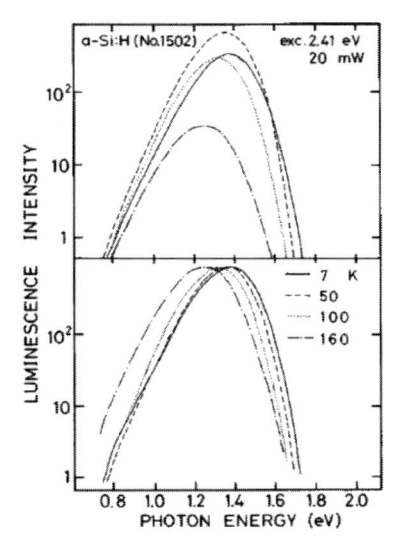

Figure 3.23 PL spectra measured at various temperatures for a-Si:H (the upper figure) and those normalized at the PL peak intensity (the lower figure). [Reproduced from Yamaguchi and Morigaki, *Phys. Rev. B*, **55**, 2368 (1997) by permission of American Physical Society.]

Figure 3.24 is a schematic illustration of the PL spectra in a-Si:H films. The PL spectra in a-Si:H consist of two bands, that is, principal PL and defect PL. The former has a peak at 1.3–1.4 eV with the FWHM of 0.3 eV. The latter is of the energy below 0.9 eV and appears in a-Si:H films of high defect density.

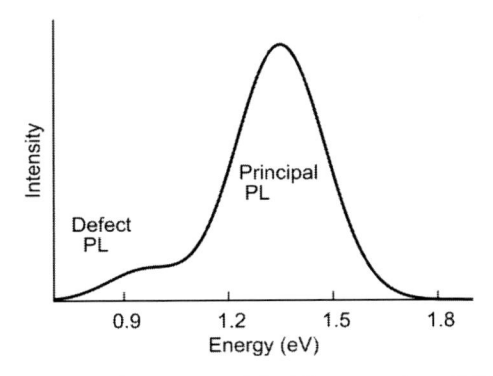

Figure 3.24 Schematic illustration of the PL spectra in a-Si:H films.

The interpretation of the PL in a-Si:H still contains controversial issues. Dunstan and Boulitrop [1984] have shown that most of the aspects of the principal PL can be described by a model considering exponential tail without further assumptions such as electron–phonon coupling. However, some experimental results reported later, for example, PA and ODMR in a-Si:H/a-Si$_3$N$_4$:H multilayers and optically detected ENDOR (ODENDOR) in a-Si:H, suggest the self-trapping of the holes in a-Si:H at low temperatures [Morigaki, 1992], as mentioned in Section 3.2.3.

The authors have proposed a model in which the principal PL is dominated by recombination of electrons at tail states and self-trapped holes at low temperatures. In addition, the PL due to excitonic state has also been shown by ODMR measurements and lifetime distributions obtained by means of frequency resolved spectroscopy (see Section 3.3.3). At the temperature higher than nearly 100 K, the recombination of electrons at the tail states of the conduction band and holes at the tail states of the valence band dominates the principal PL.

The intensity of the PL decreases with raising temperature above nearly 100 K [Engemann and Fischer, 1974]. This decrease has been attributed to the thermal excitation of electrons to the conduction band followed by nonradiative recombination. In this case, temperature variation of the PL intensity is described by

$$I(T) \propto \frac{p_r}{p_r + p_0 \exp\left(-\varepsilon/k_B T\right)}. \tag{3.12}$$

The temperature variation of the PL intensity commonly observed in a-Si:H is of the form described by,

$$I(T) \propto [1 + A\exp(T/T_0)]^{-1}. \tag{3.13}$$

This relation is derived from Eq. (3.12) by assuming an appropriate distribution of the activation energy as has been reported by Collins and Paul [1982].

3.3.3 Defects and Recombination Processes through Gap States

In the case of defective a-Si:H, creation of gap states by the defects gives rise to additional recombination processes through the gap states. The recombination processes are either radiative or nonradiative. The intensity of principal PL in a-Si:H decreases with increasing defect density because the Si-dangling bonds, which are dominant defects in a-Si:H, act as nonradiative recombination centers. The nonradiative recombination through the Si-dangling bonds competes with radiative recombination, as has been shown by the quenching signal observed in ODMR measurements. On the contrary, observation of defect PL indicates the existence of radiative defects other than the Si-dangling bonds. The radiative defects may be related to nitrogen incorporated in the a-Si:H films as impurity [Yamaguchi *et al.*, 1989, 1991].

When the radiative and nonradiative recombination compete together, the intensity of the PL, I, is given by

$$I \propto \eta = \frac{P_{\mathrm{R}}}{P_{\mathrm{R}} + P_{\mathrm{NR}}}, \tag{3.14}$$

where, η, P_{R}, and P_{NR} denote the quantum efficiency, the radiative, and nonradiative recombination rates, respectively.

The nonradiative recombination at low temperatures has been attributed to tunneling of the electron from the tail state to the closest dangling bond. In this case, the nonradiative recombination rate strongly depends on the separation between the electron and the dangling bond, R, as described by [Street *et al.*, 1978]

$$P_{\mathrm{NR}} = \omega_0 \exp\left(-2R/R_0\right) \tag{3.15}$$

where ω_0 and R_0 denote a frequency of the order of 10^{12} s^{-1} and the radius of the wavefunction of the tail state occupied by the electron before the tunneling, respectively.

Dependence of quantum efficiency, η, on the defect density, N_{D}, has been discussed by Street *et al.* [1978] as follows. When R is smaller than a distance called critical transfer radius defined by

$$R_{\mathrm{c}} = R_0 \log(\omega_0/P_{\mathrm{R}}), \tag{3.16}$$

the nonradiative recombination is more probable than radiative recombination of the rate of P_{R}. As P_{NR} exponentially depends on R, the recombination is radiative ($P_{\mathrm{R}} \gg P_{\mathrm{NR}}$) when $R > R_{\mathrm{c}}$, and nonradiative ($P_{\mathrm{NR}} \gg P_{\mathrm{R}}$) when $R < R_{\mathrm{c}}$. Therefore, η is proportional to the probability of finding no dangling bond within the distance of R_c. In the case of random spatial distribution of the dangling bonds, η is obtained by

$$\eta = \exp\left(-\frac{4\pi R_c^3 N_D}{3}\right). \tag{3.17}$$

showing that the PL intensity exponentially depends on the defect density. The experimental results of the PL intensities in a-Si:H with various defect densities have been fitted well by the above equation [Street *et al.*, 1978].

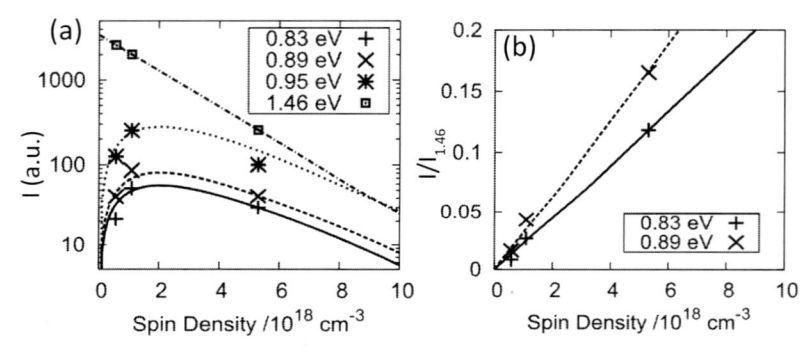

Figure 3.25 (a) Intensities, *I*, of the principal PL and defect PL and (b) ratio of the intensity of defect PL, *I*, to that of principal PL of 1.46 eV, $I_{1.46}$, plotted as functions of defect density N_D.

The intensities of the defect PL in a-Si:H with various defect densities have also been reported. The results for the a-Si:H films as grown, shown in Fig. 3.25, have been discussed on the basis of the above discussion as follows [Ogihara *et al.*, 2006]. In the case of the defect PL, the intensity, I_D, is proportional to the density of the radiative defects, N_R, and the quantum efficiency, that is,

$$I_D \propto N_R \eta. \tag{3.18}$$

The quantum efficiency for the defect PL is also affected by the density of nonradiative defects (Si-dangling bonds) at which competing nonradiative recombination occurs and may be expressed by Eq. (3.17). Assuming that $N_R \propto N_D$, we obtain

$$I_D \propto N_D \exp\left(-\frac{4\pi R_c^3 N_D}{3}\right). \tag{3.19}$$

The intensity of the principal PL, which is proportional to η given by Eq. (3.17), is described by,

$$I_{PR} \propto \exp\left(-\frac{4\pi R_c^3 N_D}{3}\right). \tag{3.20}$$

The curves described by Eqs. (3.19) and (3.20) are shown in Fig. 3.25(a). Equation (3.19) well describes the intensity of the defect PL in the films as grown, as shown in Fig. 3.25(a). Equation (3.19) also predicts that the strongest intensity of the defect PL is observed when $N_D = 3/(4\pi R_c^3)$. Figure 3.25(b) shows that the results are fitted by $I_D/I_{PR} \propto N_D$ expected from Eqs. (3.19) and (3.20).

The above results clearly show that there are at least two recombination processes through gap states in a-Si:H. One is nonradiative recombination through Si-dangling bonds and the other is recombination at radiative defects responsible for the defect PL.

3.3.4 Lifetime Distribution of the PL in a-Si:H

Nonexponential decays of the PL are generally observed in a-Si:H and understood by considering lifetime distributions. The decay of the PL, $I(t)$, is described by

$$I(t) \propto \int_0^\infty P(\tau)\tau^{-1}\exp(-t/\tau)d\tau, \tag{3.21}$$

where τ and $P(\tau)$ denote the lifetime and its distribution, respectively. In the case of broad lifetime distributions, they are often represented by plots of $P'(\log\tau)$ vs. $\log\tau$, where P' is defined by $P'(\log\tau) = \tau P(\tau)$ so that $P'(\log\tau)d(\log\tau) = P(\tau)d\tau$ and Eq. (3.21) is rewritten as

$$I(t) \propto \int_{-\infty}^\infty P'(\log\tau)\tau^{-1}\exp(-t/\tau)d(\log\tau). \tag{3.22}$$

The lifetime distributions can be obtained from the PL decays measured by means of time-resolved spectroscopy (TRS). However, another technique called frequency-resolved spectroscopy (FRS) is more useful to obtain the lifetime distributions because a lock-in amplifier used in this technique improves the signal-to-noise ratio and is suitable to measure the long lifetime components whose signals are normally weak in TRS measurements. FRS has been applied to the PL in a-Si:H by Depinna and Dunstan [1984] for the first time. In the application of FRS in PL measurements, the excitation intensity is modulated at an angular frequency, ω. The lock-in amplifier is set in-phase or quadrature (out of phase) by monitoring the scattered excitation light before measuring the PL. For the quadrature FRS (QFRS), the phase of the lock-in amplifier is set so that the output for the signal of the scattered excitation light is zero and the out-of-phase signal of the PL, I_y, is measured. For in-phase FRS, the lock-in amplifier is set to maximize the output for the signal of the scattered excitation light and the in-phase signal of the PL, I_x, is measured. (In order to correct the instrumental response, both I_x and I_y must be obtained for the PL together with those of the scattered excitation. In this case, the measurements can be done without adjustment of the phase because the phase can be corrected later together with the instrumental response.)

The plots of I_y vs. $-\log\omega$ obtained by QFRS measurements directly represent the lifetime distributions, that is, plots of $P'(\log\tau)$ vs. $\log\tau$ with $\tau = \omega^{-1}$. A rigorous discussion of the interpretation of the FRS signals has been given by Stachowitz *et al.* [1994]. QFRS measurements of PL in a-Si:H have been done by many research groups and considerable information on the nature of the recombination have been obtained.

The in-phase signal, I_x, plotted as a function of ω also represents the information of the lifetime distribution. The relation between I_x and $P(\tau)$ is approximately described by

$$I_x \propto \int_0^{1/\omega} P(\tau)\mathrm{d}\tau = \int_{-\infty}^{-\log\omega} P'(\log\tau)\mathrm{d}(\log\tau). \tag{3.23}$$

The plot of I_x vs. $-\log\omega$ can be used to obtain a characteristic value of the lifetime as described later.

The lifetime distribution of PL in a-Si:H is schematically illustrated in Fig. 3.26. The radiative recombination rate of an electron–hole pair, P_R, strongly depends on the separation between the electron and the hole, r, as described by

$$P_R = \omega_R \exp\left(-\frac{2r}{a}\right). \tag{3.24}$$

where $\omega_R \sim 10^8$ s^{-1} and a denote the radius of more extended wave-function of the carriers (see, e.g., Tsang and Street, 1979). The distribution of r gives rise to a broad featureless lifetime distribution.

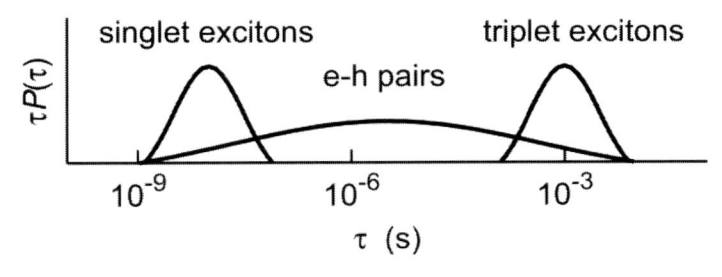

Figure 3.26 Schematic illustration of lifetime distribution of the PL in a-Si:H.

As P_R increases exponentially with decreasing r, an electron normally recombines with the closest hole. A pair of an electron and a hole created simultaneously by the same photon is called a geminate pair (see Section 1.3). In the case of low generation rate, an electron and the closest hole are normally a geminate pair. In this case, the recombining pairs are geminate pairs and the recombination is called geminate recombination. Separation of an electron and a hole of a geminate pair is independent of generation rate of the carriers. Therefore, the lifetime distribution does not depend on the intensity of the excitation light when the excitation intensity is weak and geminate recombination occurs. In the case of the high generation rate, the electron and the closest hole are not always created by the same photon, and the recombining electron–hole pair is not a geminate pair. In such a case, the recombination is called nongeminate recombination and the electron–hole pairs are called nongeminate pairs or distant pairs. With increasing generation rate, the average separation of the electron and hole of the nongeminate pair decreases and the lifetime distribution shifts toward shorter lifetimes. The broad featureless lifetime distribution and its generation rate dependence have been observed in FRS measurements of the PL in a-Si:H.

In addition, the emission from triplet states also contributes to the PL in a-Si:H as has been confirmed by ODMR for the PL at the delay time of nearly 1 ms after the excitation is turned off [Yoshida *et al.*, 1989; Yoshida and Morigaki, 1989]. A narrow peak at 1 ms has been observed in the lifetime distribution of the PL in a-Si:H and has been attributed to emission from triplet excitons [Ogihara, 1998; Aoki *et al.*, 2002].

Normally, the lifetime of singlet excitons is expected to be nearly 10 ns. The peak at nearly 10 ns in the lifetime distribution of the PL in a-Si:H is seen at the energies higher than 1.7 eV [Ogihara *et al.*, 2000; Takemura *et al.*, 2002]. At lower energies including the peak of the PL spectra, another component at microsecond region dominates the lifetime distribution and the nanosecond component is not seen apparently in most of the experiments. On the basis of the results without nanosecond components, some research groups have attributed the microsecond component to emission from singlet excitons [e.g., Aoki *et al.*, 2002]. However, the origin of the PL that dominates the microsecond component is not pure singlet state as has been suggested by ODMR experiments. In the ODMR measurements in a-Si:H, an enhancing signal of electron–hole pairs, which should not be observed for the singlet excitons, is seen for the PL having lifetimes of microsecond region. The enhancing signal is observed in the case of the electron–hole pairs whose Zeeman levels are described by $|\uparrow\uparrow\rangle$, $|\downarrow\downarrow\rangle$, $|\uparrow\downarrow\rangle$, and $|\downarrow\uparrow\rangle$, where the arrows denote the directions of the spins of the electron and the hole. Thus, the PL is dominated by the recombination of electron–hole pairs, although the excitonic PL actually contributes to the PL in a-Si:H.

In the case of a-Si$_{1-x}$N$_x$:H alloys, the nanosecond peak in the lifetime distribution of PL becomes dominant with increasing x [Takemura *et al.*, 2002], showing that the PL is dominated by excitonic PL. The Coulomb interaction between the electrons and holes in the a-Si$_{1-x}$N$_x$:H alloys is stronger than that in a-Si:H, as indicated by difference of the refractive indices. As a result, the formation of the excitons in a-Si$_{1-x}$N$_x$:H alloys is more probable than that in a-Si:H.

3.3.5 Recombination Rates of the Electron–Hole Pairs in a-Si:H

Lifetime distribution of the electron–hole pairs is broad and featureless. This is due to broad distribution of radiative and nonradiative recombination rates. In order to analyze the recombination rates quantitatively, determination of single values that characterize the distributions is desired. For example, obtaining the peak position, the average, and the intermediate value in the distribution enable us to see the temperature dependence by plotting those as functions of temperature. In the case of electron–hole pairs in a-Si:H, it is difficult to determine the peak position of the lifetime distribution accurately. However, the intermediate value can be determined easily and directly from the in-phase FRS as follows.

As expected from Eq. (3.23), I_x is proportional to the number of the electron–hole pairs having lifetime shorter than ω^{-1}. When ω is sufficiently low, all of the electron–hole pairs responsible for the PL can contribute to I_x. Therefore, I_x increases with decreasing ω and finally reaches a saturated value, I_s. An angular frequency $\omega_{1/2}$ that satisfies $I_x(\omega_{1/2}) = \frac{I_s}{2}$ can be obtained from I_x vs. ω plot. A lifetime $\tau_{1/2} = \omega_{1/2}^{-1}$ is considered to be intermediate lifetime, as I_x/I_s is the ratio of the number of electron–hole pairs having lifetime shorter than ω^{-1} to the total number of the electron–hole pairs responsible for the PL.

The characteristic values of the radiative and nonradiative recombination rates can also be estimated by applying relations between the recombination rates and the lifetime without distributions as follows. When the radiative and nonradiative recombination rates have no distribution, the quantum efficiency is given by Eq. (3.14) and the lifetime by

$$\tau = \frac{1}{P_R + P_{NR}}. \tag{3.25}$$

From Eqs. (3.14) and (3.25), we obtain

$$P_R = \eta\tau^{-1} \tag{3.26}$$

and

$$P_{NR} = \tau^{-1} - P_R.$$ (3.27)

Thus, the rates of the radiative and nonradiative recombination processes competing together are obtained by measuring the lifetime and the intensity of the PL. When the recombination rates have broad distributions, characteristic values of the recombination rates can be obtained by substituting τ by $\tau_{1/2}$ and estimating the quantum efficiency from I_s [Ogihara *et al.*, 2008].

Temperature variations of P_R, and P_{NR} have been obtained for PL in a-Si:H films by applying the method described above. Increase of P_R with raising temperature has been observed for the recombination of electron–hole pairs responsible for the principal PL and defect PL in a-Si:H. An example of the results of P_R for the defect PL is shown in Fig. 3.27 [Ogihara *et al.*, 2011]. The increase of the radiative recombination rate has been attributed to thermal excitation of carriers to shallow tail states. The shallow tail states are considered to be more extended than deep tail states or self-trapped holes that dominate the PL at low temperatures. The radius of the wavefunction of the more

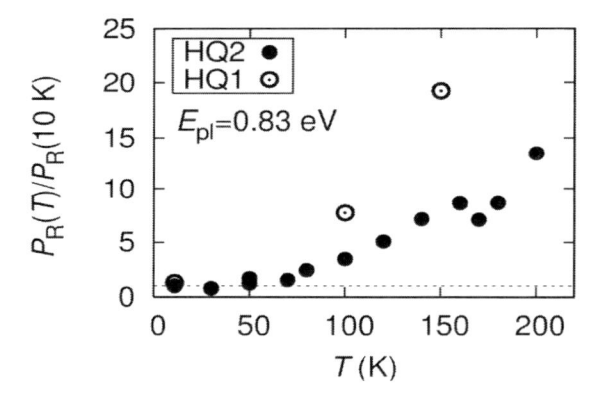

Figure 3.27 Temperature variation of the radiative recombination rate of the electron–hole pairs responsible for the defect PL in a-Si:H. [Reproduced from Ogihara *et al.*, *Phys. Status Solidi C*, **8**, 2792 (2011) by permission of John Wiley and Sons.] The data for the films HQ1 and HQ2 were obtained after illumination of pulsed light of 2.48 eV and 1.55 eV, respectively.

extended carrier, a, affects P_R as expected from Eq. (3.24). In the case of principal PL due to recombination of the tail electrons and self-trapped holes, a is the radius of the electrons. Thus, increase of the contribution of electrons at shallow tail states with rising temperature gives rise to increase of P_R. In the case of defect PL, the temperature variation of P_R, which is different from that for the principal PL, has been attributed to the change of the hole's electronic states [Ogihara et al., 2011].

Temperature dependence of the nonradiative recombination rate has also been estimated for the case of defect PL [Ogihara et al., 2012]. The experimental results are fitted by a relation predicted by Englman and Jortner [1970] for the case of strong electron–phonon coupling. The nonradiative recombination process for the case of strong electron–phonon coupling is described by configuration coordinate model as shown in Fig. 3.28. In this case, the temperature dependence of P_{NR} is expressed by

$$P_{NR} = AT^{*-1/2}\exp\left(-\frac{\varepsilon}{k_B T^*}\right) \tag{3.28}$$

and

$$T^* = \frac{\hbar\Omega_0}{2k_B}\coth\left[\left(\frac{\hbar\Omega_0}{2k_B}\right)\frac{1}{T}\right], \tag{3.29}$$

where A, T^*, Ω_0, and ε denote a factor, the effective temperature, the frequency of the phonon, and the activation energy, respectively. The factor A is affected by the radius of the electron and the separation between the electron and the nonradiative center similarly to Eq. (3.15). The factor A is independent of the temperature when the electrons are at gap states created by the radiative defects irrespective of the temperature. At high temperatures, $T^* \simeq T$ and $P_{NR} \propto T^{-1/2} \exp(-\varepsilon/k_B T)$. As $T^{-1/2}$ varies with T much more slowly than $\exp(-\varepsilon/k_B T)$, the temperature dependence of P_{NR} is like that of activation form at high temperatures. At low temperatures, T^* is almost independent of T, so P_{NR} becomes independent of the temperature.

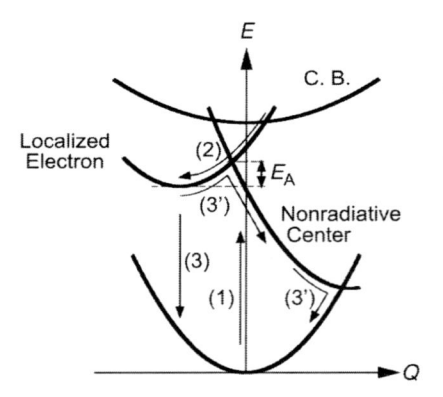

Figure 3.28 Schematic diagram of the configuration coordinate model in the case of strong electron–phonon coupling. (1) Optical excitation, (2) thermalization of the electron, (3) radiative recombination, and (3′) nonradiative recombination.

For the case of weak electron–phonon coupling, multiphonon emission occurs between two curves shown in Fig. 3.29. The number of the emitted phonons is approximately $\frac{\Delta E}{\hbar \Omega_M}$, where ΔE and Ω_M denote the difference of energy shown in Fig. 3.29 and the highest frequency of the phonons, respectively.

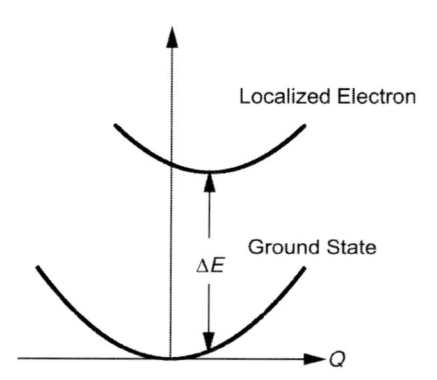

Figure 3.29 Schematic diagram of the configuration coordinate model in the case of weak electron–phonon coupling.

In this case, P_{NR} is independent of the temperature as described by

$$P_{NR} = W_0 \exp\left(\frac{-\gamma' \Delta E}{\hbar \Omega_M}\right),$$
(3.30)

where γ' is a constant of the order of 1. This case is not fitted to the experimental results for the defect PL.

Temperature dependence of P_{NR} is also expected for the case of nonradiative recombination that follows thermal excitation of the electron to the conduction band. Equation (3.12), which describes the intensity of the principal PL, is based on this model. However, the experimental results for the defect PL with constant P_{NR} at low temperatures cannot be explained by this model. A detailed discussion has been done by the authors [Ogihara *et al.*, 2012].

3.4   Light-Induced Phenomena in a-Si:H

3.4.1   Introduction

Metastable change of dark and photo-conductivity caused by illumination of light in a-Si:H was discovered by Staebler and Wronski in 1977. The changes of the conductivities have been understood as results of light-induced creation of Si-dangling bonds, which has been confirmed by ESR measurements (see Section 3.5.1) and optical absorption at the energy in the subgap region. The changes caused by illumination are recovered after annealing at 160–200°C.

The light-induced phenomena in a-Si:H are generally called the Staebler–Wronski effect. The Staebler–Wronski effect is an interesting phenomenon in physics of amorphous semiconductors. In addition, it is important to understand and suppress the Staebler–Wronski effect in application of a-Si:H to devices such as solar cells in which the creation of dangling bonds causes the degradation of performance.

In this section, experimental results related to the light-induced defect creation in a-Si:H are described except those of ESR that are shown in Section 3.5 with models describing the mechanism of light-induced creation of Si-dangling bonds.

3.4.2 Light-Induced Effects in Conductivity

Staebler and Wronski have found that the photoconductivity decreases by one order of magnitude during illumination of intense light as shown in Fig. 3.30.

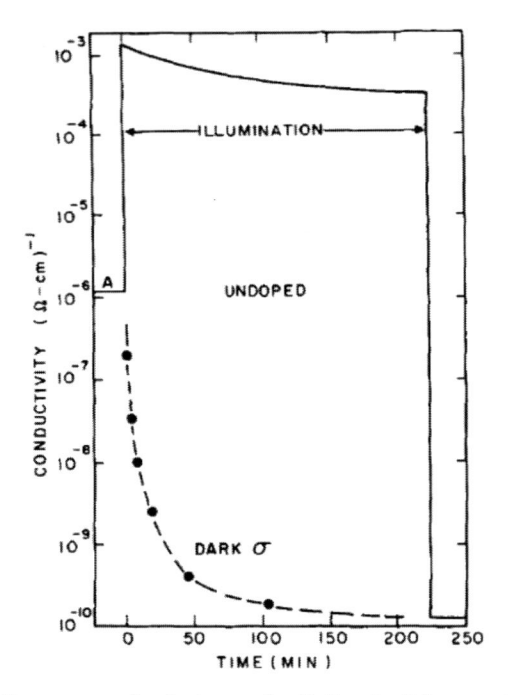

Figure 3.30 Decrease of photo-conductivity (solid curve) and dark conductivity during illumination with 200 mW/cm^{-2} of filtered tungsten light (600–900 nm) for an undoped a-Si:H. [Reproduced from Staebler, D. L. and Wronski, C. R., *J. Appl. Phys.*, **51**, 3262 (1980) by permission of American Institute of Physics.]

At the same time, dark conductivity also decreases by four orders of magnitude, as also shown in Fig. 3.30.

These changes are understood as results of light-induced creation of defects. Creation of gap states causes the decrease of Fermi energy, increase of the activation energy, and decrease of the dark conductivity. The decrease of photo conductivity is attributed to decrease of lifetime caused by recombination at photo-created defects.

3.4.3 Optical Absorption

Light-induced creation of the defects gives rise to increase of the optical absorption at the energies in subgap region. Normally, a-Si:H is prepared as a thin film, so that small absorption coefficient in subgap region is difficult to obtain by measurements of transmittance spectra. Measurements of the small absorption coefficient in a-Si:H have been done by using techniques for improving sensitivity such as photo-thermal deflection spectroscopy (PDS) and constant photocurrent method (CPM). Figure 3.31 shows the change of the absorption spectrum in a-Si:H by illumination and recovery after annealing. The increase of absorption coefficient is due to gap states created by light-induced defects.

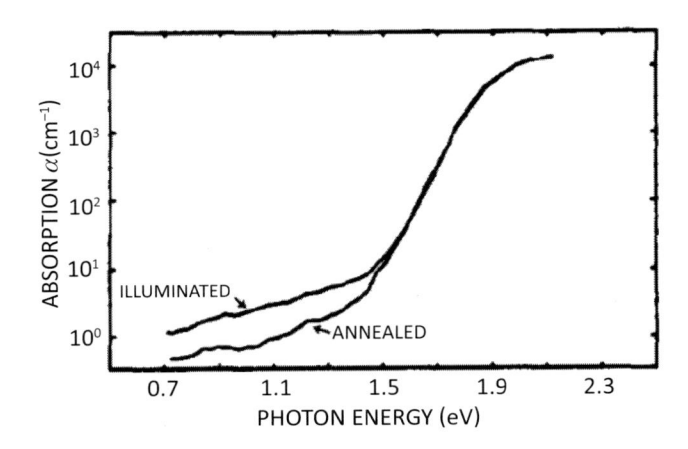

Figure 3.31 The effect of the illumination on the absorption spectrum of undoped a-Si:H. [Reproduced from Amer *et al.*, *Physica*, **117 B**, 897 (1983) by permission of Elsevier.]

3.4.4 Photoluminescence

Decrease of the intensity of the principal PL in a-Si:H after prolonged illumination, which is called PL fatigue, has been observed [Morigaki *et al.*, 1980; Pankove and Berkeyheiser, 1980]. An example of the results is shown in Fig. 3.32. The PL intensity also recovers after annealing. As mentioned in Section 3.3, Si-dangling bonds in a-Si:H act as nonradiative recombination centers. Therefore, the PL fatigue is understood as a result of nonradiative recombination through photo-created dangling bonds. The lifetime of the principal PL also decreases after illumination, indicating the increase of the nonradiative recombination rate. On the contrary, defect PL increases in intensity after illumination as shown in Fig. 3.32 [Pankove and Berkeyheiser, 1980], indicating that the illumination creates the radiative defects as well.

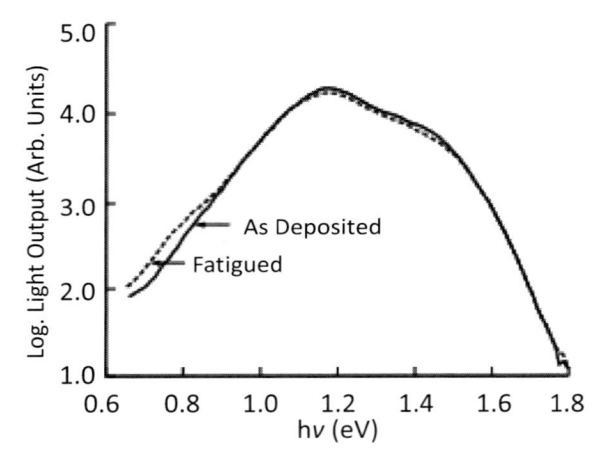

Figure 3.32 Photoluminescence spectrum of a-Si:H at 78 K before and after 10 min irradiation at 5 W/cm. [Reproduced from Pankove and Berkeyheiser, *Appl. Phys. Lett.*, **37**, 705 (1980) by permission of American Institute of Physics.]

Assuming that the number of photo-created radiative defects is proportional to that of photo-created dangling bonds, the intensity of the defect PL is predicted by Eq. (3.20) and its change with increasing defect density is expected to be similar to the curves shown in Fig. 3.25(a). However, intensities of the defect PL larger than the maximum shown in Fig. 3.25(a) have been observed for some a-Si:H films after illumination [Ogihara *et al.*, 2006]. Thus, the experimental results of the intensity of the defect PL as functions of the number of photo-created defects are not fitted by Eq. (3.20), indicating that the radiative recombination at photo-created radiative defects is not affected strongly by the photo-created Si-dangling bonds. The discrepancy between Eq. (3.20) and the experimental results for a-Si:H films after illumination have been understood by considering inhomogeneous spatial distributions of photo-created defects [Ogihara *et al.*, 2006].

The difference between recombination processes at the photo-created defects and native defects in a-Si:H has been studied by obtaining the temperature variations of the recombination rates by the method described in Section 3.3. However, the experimental results up to the present have been understood without considering difference of recombination processes through the photo-created defects and the native defects [Ogihara *et al.*, 2011].

3.5 Light-Induced Defect Creation in a-Si:H

3.5.1 Introduction

The first evidence for light-induced creation of Si-dangling bond under optical excitation by a mercury lamp of 500 W has been obtained from the ESR measurements by Hirabayashi *et al.* [1980], using an a-Si:H film of 1 µm thick prepared at 300°C by PECVD. Two components of $g = 2.0046$ and $g = 2.013$ were observed, as shown in Fig. 3.33(a), while after optical excitation, a line of $g = 2.005$ was still observed that was identified as due to Si-dangling bonds, as shown in Fig. 3.33(b). Hirabayashi *et al.* [1980] suggested that a dangling bond is created as a result of breaking of a weak Si–Si bond adjacent to a-Si–H bond under optical excitation. They also suggested that as strength of the Si–H bond is greater than the Si–Si bond's one, electronegativity of hydrogen is larger than that of silicon, so that the Si–Si bond adjacent to the Si–H bond becomes weak. After then, Dersch *et al.* [1981] independently made a detailed ESR measurement on a powder (small pieces) sample of a-Si:H prepared at 280°C by PECVD, in which the ESR signal of $g = 2.0055$ is increased after illumination with focused light of 100 W halogen lamp for 15 h compared with that before illumination, as shown in Figs. 3.34(a) and (b), and they confirmed that this signal is annealed out at 220°C, as shown in Figs. 3.34(c) and (d). As seen in the figure, they obtained an ESR signal with better S/N ratio than that shown in Fig. 3.33, using a powder sample. They suggested that the neighboring Si–H bond is switched toward a broken bond, leaving a dangling bond, and then the dangling bond moves at sufficiently high temperature by a continuous exchange of hydrogen atoms.

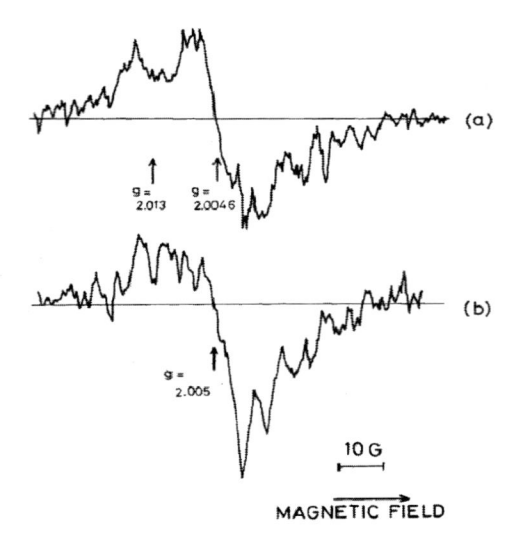

MAGNETIC FIELD

Figure 3.33 ESR spectra of an a-Si:H film taken at 77 K and microwave frequency of 9.2 GHz and microwave power of 5 mW, (a) under optical excitation by a mercury lamp of 500 W, and (b) in the dark just after the excitation. These spectra were obtained by subtracting the background associated with the glass substrate by use of a computer. Before the optical excitation, any ESR signal associated with the sample was not observed in the dark at 77 K. [Reproduced from Hirabayashi *et al.*, *Jpn. J. Appl. Phys.*, **19**, L357 (1980) by permission of The Japan Society of Applied Physics.]

For comparison, an ESR signal due to dangling bonds in a high-quality a-Si:H film prepared at 250°C taken after intense pulsed illumination for 2 h [a YAG optical parametric oscillator system (OPO), operating at 1.55 eV] is shown in Fig. 3.35, in which a number of dangling bonds are created (Morigaki *et al.* [2003]; for detail, see Section 3.5.7).

The dangling bond has been concluded to act as a nonradiative recombination center as a result of luminescence fatigue in a-Si:H by Hirabayashi *et al.* [1980] and Pankove and Berkeyheiser [1980]. Pankove and Berkeyheiser [1980] also suggest that the weak Si–Si bond without neighboring hydrogen, that is, the normal weak Si–Si bond, is broken by illumination. This is in contrast with the model of Hirabayashi *et al.* [1980] that the weak Si–Si bond involving the Si–H bond plays an important role in the light-induced defect creation in a-Si:H as mentioned above.

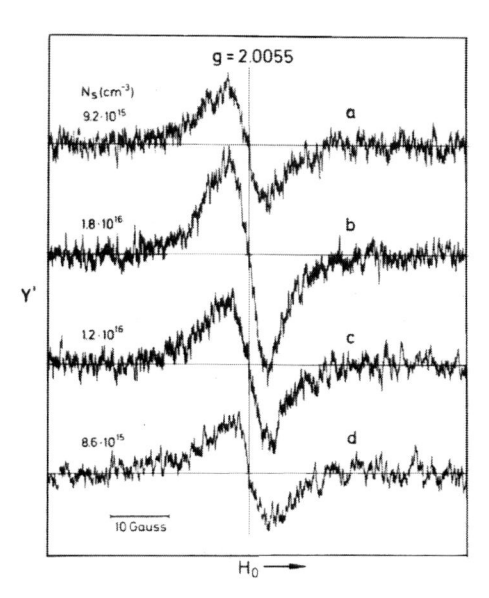

Figure 3.34 ESR spectra of an a-Si:H powder sample measured at 20°C in the X band (a) after annealing at 220°C for 1 h, (b) after illumination with focused light of 100 W halogen lamp for 15 h, (c) after annealing at 220°C for 30 min and (d) subsequent annealing at 220°C for additional 30 min. [Reproduced from Dersch *et al.*, *Appl. Phys. Lett.*, **38**, 456 (1981) by permission of American Institute of Physics.]

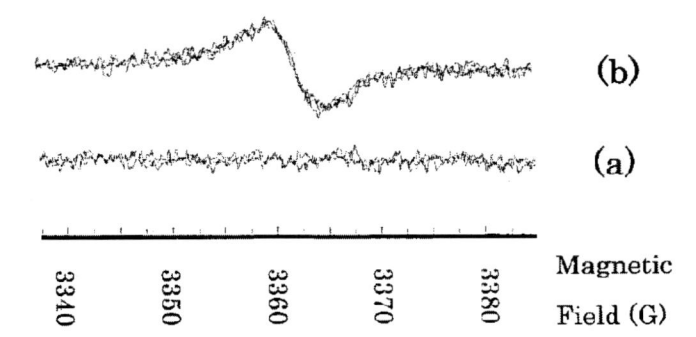

Figure 3.35 ESR spectra (three traces) observed at room temperature and 9.4 GHz for a 1 μm thick a-Si:H film, (a) before pulsed illumination at 1.55 eV (800 nm), (b) after 2 h pulsed illumination (total illumination time = 0.79 ms) at 1.55 eV (800 nm). [Reproduced from Morigaki *et al.*, *Philos. Mag. Lett.*, **83**, 341 (2003) by permission of Taylor & Francis.]

There are experimental facts suggesting that hydrogen is involved in the processes of light-induced defect creation in a-Si:H. Vignoli *et al.* [1996] have shown a correlation between the defect creation efficiency C_d and the structural factor R, as shown in Fig. 3.36. C_d is defined by a proportionality coefficient with which the defect-creation rate is proportional to the product of

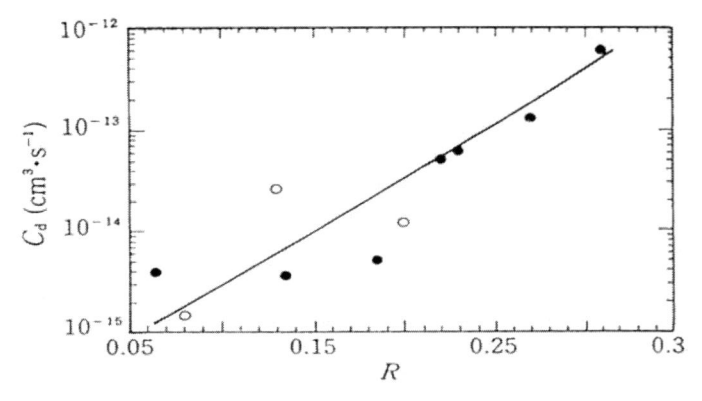

Figure 3.36 C_d *vs.* R (see the text). The solid circles are cited from [Caputo *et al.*, 1994; Vignoli *et al.*, 1996].

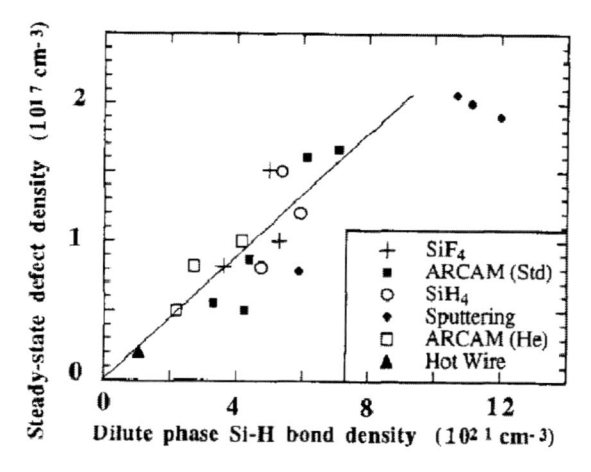

Figure 3.37 Steady-state defect density under high-intensity illumination as a function of the dilute phase Si–H bond density. The data represented by +, ●, O, and ◆ were obtained by light soaking under Kr⁺ laser at RT by the Princeton group. The line is drawn as a guide for the eyes [Roca i Cabarrocas *et al.*, 1989; Godet *et al.*, 1996].

free electron density and free hole density. R is defined by the total hydrogen content excluding the monohydride (Si–H bond) content relative to the total hydrogen content. For a-Si:H samples containing a lot of Si–H$_2$ bonds, Si–H$_3$ bonds, (Si–H)$_n$ bonds, and so on, it has been found that a number of dangling bonds are created under illumination. There is also some evidence that monohydride also plays an important role in the light-induced defect creation in a-Si:H [Godet *et al.*, 1996], as shown in Fig. 3.37, in which correlation between steady-state dangling bond density and dilute phase Si–H bond density is shown.

3.5.2 Mechanism for Light-Induced Defect Creation in a-Si:H

The models of light-induced defect creation in a-Si:H are classified as (1) weak Si–Si bond breaking model, (2) hydrogen-related bond (Si–H, Si–H–Si) breaking model, (3) hydrogen-mediated model, (4) charged center-trapping model, and (5) impurity-involved model, as follows:

(1) The weak Si–Si bond breaking model

This model has been considered by Stutzmann *et al.* [1985] in detail. A weak Si–Si bond adjacent to a Si–H bond is broken as a result of hole-trapping under illumination and a dangling bond is stabilized by the Si–H bond switching due to the nonradiative electron–hole recombination at the weak Si–Si bond, as shown in Figs. 3.38(a)–(c). They considered the kinetics of dangling bond density N_d as a function of illumination time t. The dangling bond is created by nonradiative electron–hole recombination, so that the following equation is given,

$$dN_d/dt = C_{SW} A_t\, np, \tag{3.31}$$

where A_t, C_{SW}, n, and p designate the recombination rate at the weak Si–Si bond, the dangling bond-creation efficiency, electron density, and hole density, respectively. Then, the dangling bond density N_d against t is approximately obtained as $N_d \propto t^{1/3}$, in agreement with the experimental result, as shown in Fig. 3.39. The dependence of N_d on the generation rate, G, follows approximately as $N_d \propto G^{2/3}$. A close agreement between theory and experiment on this dependence is obtained.

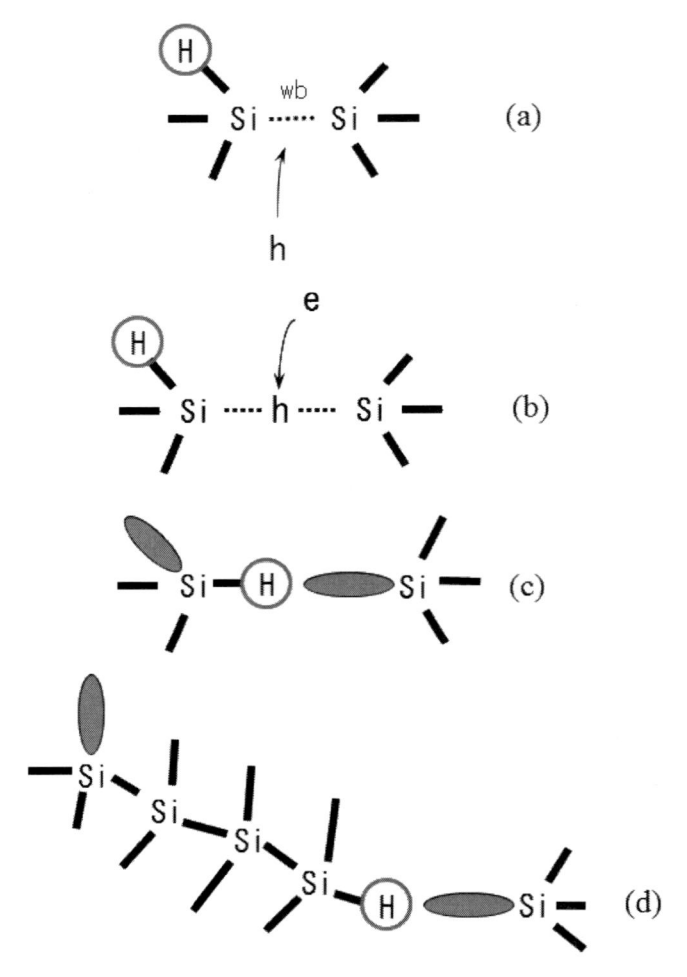

Figure 3.38 Atomic configurations involved in the formation of two types of dangling bonds, that is, normal dangling bonds and hydrogen-related dangling bonds, under illumination: (a) Self-trapping of a hole in a weak Si–Si bond (wb) adjacent to a Si–H bond, (b) Electron–hole recombination at a weak Si–Si bond, (c) Switching of a Si–H bond toward the weak Si–Si bond, leaving a dangling bond behind, and (d) Formation of two separate dangling bonds through hydrogen movement after repeating the processes shown in (a)–(c).

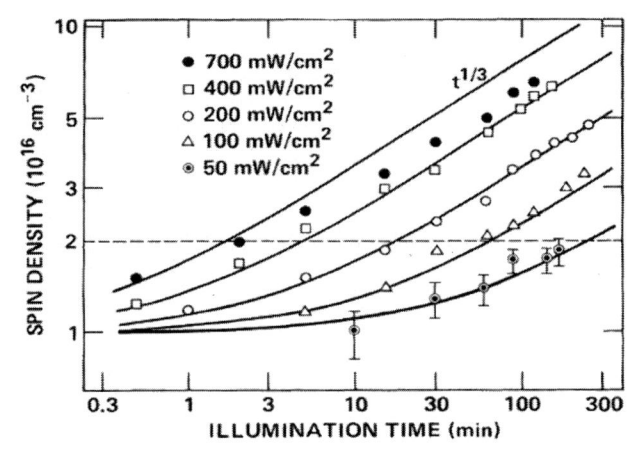

Figure 3.39 Plot of spin density N_S vs. illumination time t in a-Si:H. The solid curves represent the theoretical curve of $N_S(t) \propto t^{1/3}$ ($N_S(0) = 1 \times 10^{16}$ cm^{-3}, $C_{SW}A_t = 1.5 \times 10^{-15}$ cm^{3}s^{-1}). [Reproduced from Stutzmann *et al.*, *Phys. Rev. B*, **32**, 23 (1985) by permission of American Physical Society.]

A model [Morigaki, 1988; Morigaki and Hikita, 2007] based on the weak Si–Si bond breaking mechanism is described in more detail below: The processes involved in light-induced defect creation under illumination in a-Si:H are given as follows: A hole is self-trapped in a specific weak Si–Si bond that is a weak Si–Si bond adjacent to a Si–H bond [Fig. 3.38(a)] and then it is recombined with an electron most nonradiatively [Fig. 3.38(b)] and eventually the weak bond is broken. Using the recombination energy associated with nonradiative recombination between the electron and the hole, the Si–H bond is switched toward the weak Si–Si bond [Fig. 3.38(c)]. After switching of the Si–H bond [Fig. 3.38(c)] and breaking of the weak Si–Si bond, the two close dangling bonds are created, and further they are separated by movement of hydrogen due to hopping and/or tunneling [Fig. 3.38(d)], and eventually two separate dangling bonds [Fig. 3.38(d)], that is, a normal dangling bond and a hydrogen-related dangling bond, that is a dangling bond having hydrogen at the nearby site, are created under illumination, as shown in Fig. 3.38(d). In addition, hydrogen is dissociated from a Si–H bond located near a hydrogen-related dangling bond as a result of nonradiative recombination between an electron and a hole at the hydrogen-related dangling bond.

A dissociated hydrogen atom (metastable hydrogen atom) is inserted into a nearby weak Si–Si bond to form a hydrogen-related dangling bond. A dissociated hydrogen atom also terminates either a normal dangling bond or a hydrogen-related dangling bond. A dissociated hydrogen atom has a chance to form a hydrogen molecule by meeting together. These processes are depicted in Fig. 3.40.

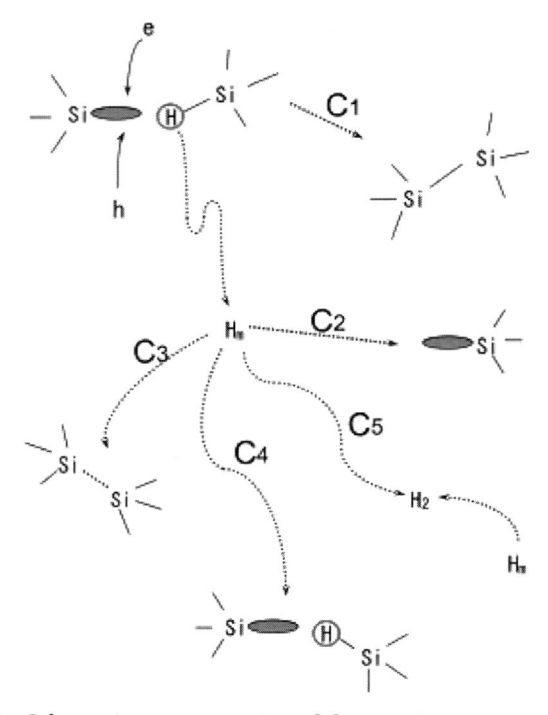

Figure 3.40 Schematic representation of the reaction processes C_1–C_5. H_m designates a dissociated metastable hydrogen atom.

The light-induced defect creation is considered in terms of a rate-equation model, taking into account the processes mentioned above. Rate equations are given as follows:

$$dN_a/dt = C_d np(N_w/N_{w0}) - C_2 N_m N_a, \tag{3.32}$$

$$dN_b/dt = C_d np(N_w/N_{w0}) - C_1 np N_b + C_3 N_m N_{Si} - C_4 N_m N_b, \tag{3.33}$$

$$dN_m/dt = C_1 np N_b - C_2 N_m N_a - C_3 N_m N_{Si} - C_4 N_m N_b - C_5 N_m^2, \quad (3.34)$$

$$dN_w/dt = -C_d np(N_w/N_{w0}) + C_2 N_m N_a + C_4 N_m N_b, \quad (3.35)$$

where N_a, N_b, N_m, N_{Si}, N_w, N_{w0}, n, and p are densities of normal dangling bonds, hydrogen-related dangling bonds, metastable hydrogen atoms, Si–Si bonds, weak Si–Si bonds adjacent to a Si–H bond, N_w at $t = 0$, free electrons and free holes including band-tail electrons and ban-tail holes, respectively, and C_d, C_1, C_2, C_3, C_4, and C_5 are reaction coefficients of the following processes. The processes of C_1, C_2, C_3, C_4, and C_5 are illustrated in Fig. 3.40: C_d is the light-induced creation of two separate dangling bonds, C_1 is the dissociation of a hydrogen atom from a Si–H bond located near a hydrogen-related dangling bond, C_2 is the termination of a normal dangling bond by a metastable hydrogen atom, C_3 is the insertion of a metastable hydrogen atom into a Si–Si bond, C_4 is the termination of a hydrogen-related dangling bond by a metastable hydrogen atom, and C_5 is the formation of a hydrogen molecule by two metastable hydrogen atoms. In Eq. (3.35), the density of weak Si–Si bonds decreases with formation of dangling bonds and increases with termination of dangling bonds by hydrogen atoms, because a Si–Si bond adjacent to a Si–H bond is considered to become a weak bond. For high-quality a-Si:H, the C_3 term is neglected, because the density of weak Si–Si bonds is relatively small compared with low-quality a-Si:H containing a large amount of hydrogen. In numerical calculations, the C_5 term is also neglected for simplicity.

First, we consider the case of relatively weak illumination such as continuous illumination [Morigaki and Hikita, 2007]. In this case, we neglect Eq. (3.35), because the weak Si–Si bond exists much more than light-induced dangling bonds, for example, $N_w \cong 10^{19} \, \text{cm}^{-3}$ and $N_d \cong 10^{17} \, \text{cm}^{-3}$, so that the density of weak Si–Si bonds under illumination is assumed not to be changed.

The carrier densities n and p are assumed to be determined by their steady-state values and by trapping of carriers (electrons) at neutral dangling bonds followed by rapid recombination with holes, and then they are proportional to G/N_d, where G and N_d are the generation rate of free carriers and the total density of neutral dangling bonds, respectively, because the trapping process at neutral dangling bonds of carriers in the conduction band and its tail states occurs faster than the kinetics of light-induced creation of dangling bonds. The steady-state values of n and p are approximately

$$n = p \cong G/(\alpha N_d), \tag{3.36}$$

where α is the trapping coefficient of free carriers by neutral dangling bonds. Then, in order to solve the rate equation numerically, the rate equations (3.32)–(3.34) are rewritten, using normalized densities of N_a, N_b, and N_m to N_{d0} [$\equiv N_d$ $(t = 0)$], that is, r, q, and s, respectively, as follows:

$$dr/dt = A/(r + q)^2 - A_2 sr, \tag{3.37}$$

$$dq/dt = A/(r + q)^2 - A_1 q/(r + q)^2 + A_3 s - A_4 sq, \tag{3.38}$$

$$ds/dt = A_1 q/(r + q)^2 - A_2 sr - A_3 s - A_4 sq - A_5 s^2, \tag{3.39}$$

$$r = N_a/N_{d0}, \tag{3.40}$$

$$q = N_b/N_{d0}, \tag{3.41}$$

$$s = N_m/N_{d0}, \tag{3.42}$$

$$N_d = N_a + N_b, \tag{3.43}$$

$$A = C_d G^2/\alpha^2 N_{d0}^3, \tag{3.44}$$

$$A_1 = C_1 G^2/\alpha^2 N_{d0}^2, \tag{3.45}$$

$$A_2 = C_2 N_{d0}, \tag{3.46}$$

$$A_3 = C_3 N_{Si}, \tag{3.47}$$

$$A_4 = C_4 N_{d0}, \tag{3.48}$$

$$A_5 = C_5 N_{d0}, \tag{3.49}$$

Second, we consider the case of intense illumination [Morigaki *et al.*, 2003; Morigaki *et al.*, 2009a]. Under this condition, the carrier densities n and p are determined by bimolecular recombination, that is, recombination between an electron and a hole created by light, and their steady-state values given below are used in rate-equations because the recombination occurs faster than the kinetics of the light-induced creation of dangling bonds.

The steady-state values of n and p are approximately

$$n = p \cong (G/\beta)^{1/2}, \tag{3.50}$$

where β is the recombination coefficient of free electrons (or tail electrons) and free holes (or tail holes), respectively. Then, the rate Eqs. (3.32)–(3.35) are rewritten to solve them numerically as follows:

$$dr/dt = B_0 w - B_2 sr, \tag{3.51}$$

$$dq/dt = B_0 w - B_1 q + B_3 s - B_4 sq, \tag{3.52}$$

$$ds/dt = B_1 q - B_2 sr - B_3 s - B_4 sq - B_5 s^2, \tag{3.53}$$

$$dw/dt = -B_0 w + B_2 sr + B_4 sq, \tag{3.54}$$

$$w = N_w/N_{d0}, \tag{3.55}$$

$$B_0 = C_d G/\beta N_{d0} w_0 \ (w_0 = w \text{ at } t = 0), \tag{3.56}$$

$$B_1 = C_1 G/\beta, \tag{3.57}$$

$$B_2 = C_2 N_{d0}, \tag{3.58}$$

$$B_3 = C_3 N_{Si}, \tag{3.59}$$

$$B_4 = C_4 N_{d0}, \tag{3.60}$$

$$B_5 = C_5 N_{d0}, \tag{3.61}$$

where r, q, s, and N_d are given by Eqs. (3.40)–(3.43). B_2, B_3, B_4, and B_5 are the same as A_2, A_3, A_4, and A_5, respectively, where A_2, A_3, A_4, and A_5 are the reaction coefficients defined in the monomolecular recombination case, where N_{d0} is N_d at $t = 0$. In the calculation for pulsed illumination, the optical excitation of pulsed light is approximated by the continuous illumination, taking the total exposure time by pulsed light as the illumination time, because the rate equations involving pulsed optical excitation are not easy to be solved numerically. This is justified within an ambiguity of a factor of about 0.74 for the calculated light-induced defect density by citing the results of Stradins *et al.* [2000] in which they measured the defect creation efficiency for pulsed illumination as a function of the separation time of two laser pulses. Thus, numerical calculations are performed, using rate Eqs. (3.51)–(3.54), under continuous illumination. The results are compared with experimental ones performed under pulsed illumination in Section 3.5.7.

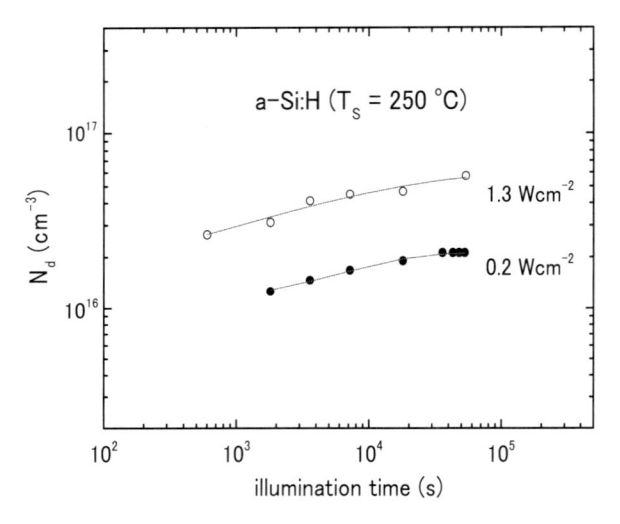

Figure 3.41 Plots of dangling bond density N_d *vs.* illumination time for illumination intensities, 0.2 and 1.3 W/cm^2 in a-Si:H (T_s = 250°C) Nos. 009151 and 601211, respectively. The curve of N_d *vs.* illumination time is fitted by the stretched exponential function given by Eq. (3.62) with $\beta = 0.809 \pm 0.055$, $\tau = (7.84 \pm 0.46) \times 10^3$ s for 0.2 W/cm^2, and $\beta = 0.403 \pm 0.057$, $\tau = (5.69 \pm 1.00) \times 10^3$ s for 1.3 W/cm^2 [Morigaki *et al.*, 2007].

The rate Eqs. (3.37)–(3.49) and (3.51)–(3.61) are solved for the two cases in the continuous illumination and the intense pulsed illumination, as will be done in Section 3.5.3 and Section 3.5.7, respectively. The experimental results on the kinetics of dangling bond density as a function of illumination time t in a-Si:H films prepared at 250°C by PECVD are shown in Fig. 3.41, in which the curves of N_d vs. t are fitted by a stretched exponential function,

$$N_d(t) = N_{ss} - [N_{ss} - N_d(0)] \exp[-(t/\tau)^\beta], \qquad (3.62)$$

where $N_d(t)$ and N_{ss} are dangling bond density at a time t after illumination is switched on and saturated dangling bond density (steady-state dangling bond density), respectively, and β and τ are a dispersive parameter and a characteristic time, respectively.

A detailed discussion on the stretched exponential relaxation will be given in the next section.

(2) Hydrogen-related bond (Si–H, Si–H–Si) breaking model

(i) *Godet*

This model [Godet, 1998] is based on an experimental fact that the light-induced dangling bond density is proportional to the density of Si–H bonds. Two Si–H bonds with hydrogen atoms adjacent to each other are transformed, as shown in Fig. 3.42(a), into two Si–H–Si bonds (bridge-bonded hydrogen BC site in crystal), which are metastable, and then the Si–H–Si bond is either captured by a Si–H bond to create a dangling bond, as shown in Fig. 3.42(b) or captured by a dangling bond to anneal out it, as shown in Fig. 3.42(c). These processes are written as reaction formula, as follows:

(a) Si–H H–Si → 2(Si–H–Si) + (Si–Si), $\qquad$ (3.63)

(b) (Si–H–Si) + Si–H → SiDB + Si–H H–Si, $\qquad$ (3.64)

(c) (Si–H–Si) + SiDB → Si–Si + Si–H. $\qquad$ (3.65)

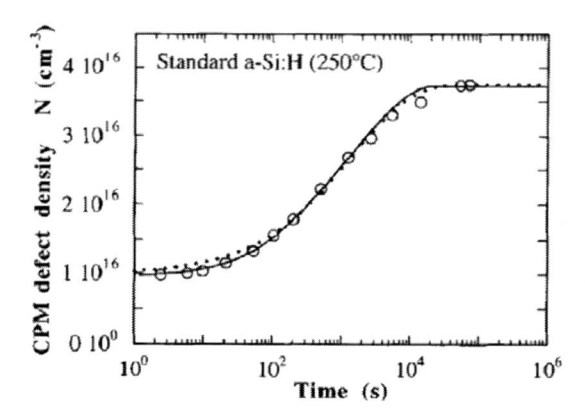

Figure 3.42 Schematic diagrams for (a): transformation of two Si-H bonds to two Si–H–Si bonds, (b) dangling bond creation from a Si–H–Si bond, and (c): annihilation of a dangling bond. [Reproduced from Godet, *Philos. Mag. B*, **77**, 765 (1998) by permission of Taylor & Francis.]

Figure 3.43 Plot of the defect density measured from CPM *vs.* illumination time in a-Si:H. The solid curve represents a calculated curve based on the model of Godet. For its detail, see Godet (1998). [Reproduced from Godet, *Philos. Mag. B*, **77**, 765 (1998) by permission of Taylor & Francis.]

Under continuous illumination, the dangling bond creation process and the dangling bond annealing process should be balanced to reach the steady state. The processes can be formulated as rate equations, and then, the kinetics of dangling bond creation can be obtained as a function of illumination time. As an example, the case of dangling bond creation under continuous illumination for a high-quality a-Si:H film prepared by PECVD is shown in Fig. 3.43, that is, the calculated curve of dangling bond density using the rate equations and its experimental values that have been measured, using the CPM. The dotted curve indicates a stretched exponential function given in Eq. (3.62) fitted to the experimental points with $N_d(0) = 9.7 \times 10^{15}$ cm^{-3}, $N_{ss} = 3.75 \times 10^{16}$ cm^{-3}, $\beta = 0.52$, and $\tau = 1.5 \times 10^3$ s.

(ii) *Branz (H-collision model)*

This model [Branz, 1999, 2003] postulates that a Si–H bond is broken with an assistance from recombination energy released from nonradiative electron–hole recombination, and eventually, a hydrogen atom is dissociated and a dangling bond is left behind. The hydrogen atom moves through the amorphous network and is inserted into a Si–Si bond to form a H–DB complex, as shown in Fig. 3.44. Two H–DB complexes collide with each other to form (Si–H)$_2$ and, as a result, metastable hydrogen atoms are annihilated. These processes are illustrated in a schematic diagram using a configuration coordinate in Fig. 3.45. On the basis of this model, Branz constructs rate equations governing the light-induced dangling bond density as a function of illumination time t and then obtains a $t^{1/3}$ dependence under weak illumination such as continuous illumination and a $t^{1/3}$ dependence under intense illumination such as pulsed illumination. Zhang and Branz [2000] considered theoretically that two electron–hole pairs (biexciton) are trapped in the Si–H bond when generated by illumination, for example, and eventually hydrogen may be released from it, as a result of nonradiative electron–hole recombination.

Figure 3.44 Schematic representation of recombination of a photo-induced electron and a photo-induced hole at an Si–H bond, subsequent dissociation of a hydrogen atom from the Si–H bond, and formation of two separate dangling bonds.

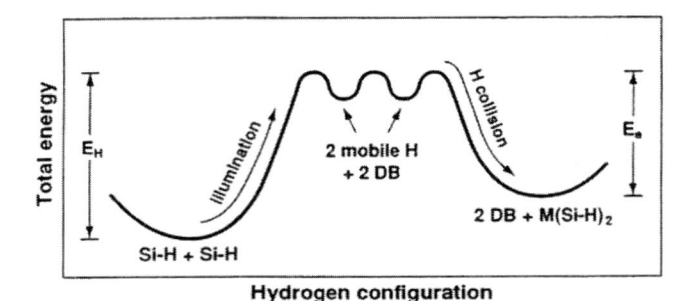

Hydrogen configuration

Figure 3.45 Configuration coordinate diagram of the reaction of defect creation. E_H and E_a represent the energies of hydrogen diffusion and metastability annealing, respectively. [Reproduced from Branz, *Philos. Mag. B*, **77**, 765 (1999) by permission of American Physical Society.]

(iii) *Longeaud* et al.

Longeaud *et al.* [2002] propose a model that hydrogen at the Si–H bond after its breaking occupies either a bond-center (BC) site or an antibonding (AB) site, because the release energy of hydrogen into a free state from a Si–H bond reaches 3.6 eV, that is too much for recombination energy, and then a dangling bond left behind is passivated by a hydrogen supplied from a hydrogen molecule at the intersite, while the other hydrogen inserts into a Si–Si bond. Further, these form hydrogen molecules leaving two dangling bonds behind, as shown in Fig. 3.46.

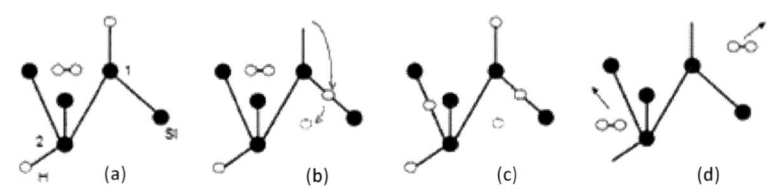

Figure 3.46 Schematic illustration of Longeaud *et al.*'s model: (a) starting configuration, (b) recombination of two carriers at a Si–H bond creates a dangling bond and a hydrogen at a BC site or an AB site, (c) this results in a breaking of the hydrogen molecule, then one hydrogen passivates the dangling bond and the other is inserted into a neighboring bond to form a hydrogen at BC site, (d) all the hydrogen atoms form hydrogen molecules, leaving two dangling bonds behind. [Reproduced from Longeaud *et al.*, *Phys. Rev. B*, **65**, 085206-1 (2002) by permission of American Physical Society.]

(iv) *Biswas and Kwon*

The configuration of a bridge-bonded hydrogen has been proposed by Biswas and Kwon [1991], as shown in Fig. 3.47(a). This configuration changes to two configurations shown in Figs. 3.47(b) and (c), that is, a strong Si–H bond and a dangling bond on the other Si atoms. Their calculation of the energetics indicates that the configuration shown in Fig. 3.47(a) corresponds to the annealed state, while the configuration shown in Figs. 3.47(b) and (c), the light-soaked state, that is a model of the Staebler–Wronski effect, according to them.

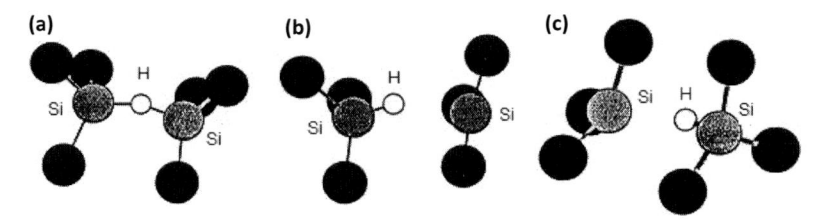

Figure 3.47 Schematic illustration: The bridge-bonded hydrogen (unshaded) interstitial defect where the hydrogen is weakly bonded to two Si atoms (shaded) in (a). In the metastable higher energy configuration (b) and (c) the hydrogen forms a strong Si–H bond with one Si atom leaving a dangling bond on the other Si. The back-bonded Si atoms are the dark circles. [Reproduced from Biswas and Kwon, *AIP Conf. Proc.* (1991) by permission of American Institute of Physics.]

(3) Hydrogen-mediated models

(i) *Zafar and Schiff*

A hydrogen-mediated model has been proposed by Zafar and Schiff [1989], in which silicon-dangling bonds are created under illumination by the transfer of hydrogen between the dilute and clustered phase of bonded hydrogen. The clustered hydrogen phase exists on the surface of microvoids, in which hydrogen is preferentially paired on sites around weak Si–Si bonds. For an example, this phase is shown in Fig. 3.48(a) as well as a dangling bond having a nearby hydrogen and paired bonded hydrogen. The dilute phase consists of two units such as shown in Fig. 3.48(b). The concentrations of two types of Si-dangling bonds represented by D_1 (dilute phase) and C_1 (clustered phase) have been obtained from elemental statistical mechanics calculations. The Si-dangling bonds with normal configuration of D_1 have been found to exist much more than those with specific configuration of C_1. This is consistent with the experimental observation for high-quality a-Si:H samples.

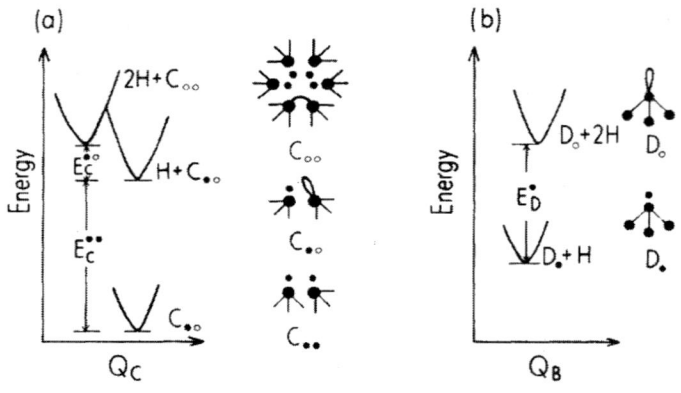

Figure 3.48 (a) Right hand side: The atomic configuration of the C center in the unhydrogenated, singly hydrogenated, and doubly hydrogenated states C_{oo}, $C_{\bullet o}$, and $C_{\bullet\bullet}$, respectively. C_1 in the text corresponds to $C_{\bullet o}$. The C_{oo} state is a weak Si–Si bond. Left hand side: Configuration coordinate diagram of the C center for three hydrogenation states. H represents an unbonded interstitial hydrogen atom. (b) Right hand side: The atomic configuration of the D center. D_o is paramagnetic and corresponds to D_1 in the text. Left hand side: The configuration coordinate diagram of the D center; the energy axes are the same as those for the C center. [Reproduced from Zafar and Schiff, *Phys. Rev. B*, **40**, 5235 (1989) by permission of American Physical Society.]

(ii) *Carlson*

The other hydrogen-mediated model associated with hydrogen motion has been proposed by Carlson [1986], in which the following items are postulated: (1) Photogenerated or injected holes are preferentially trapped on weak Si–Si bonds near the internal surfaces of microvoids [Fig. 3.49(a)]. (2) Atomic hydrogen or protons are induced to move when a hole is localized on a Si–H bond [Fig. 3.49 (b)]. (3) The atomic hydrogen or proton leaves behind a dangling bond and can move to and break a nearby weak Si–Si bond creating a new Si–H or $Si–H_2$ center and a second dangling bond [Fig. 3.49(c)]. (4) Most of the newly created dangling bonds either reconstruct on the internal surfaces or are compensated by other mobile hydrogen atoms or protons. However, a net increase in dangling bonds will occur with extensive trapping of holes as a result of incomplete bond reconstruction [Fig. 3.49(d)]. (5) Fig. 3.49(e) shows a variation of the partially reconstructed surface of Fig. 3.49(c).

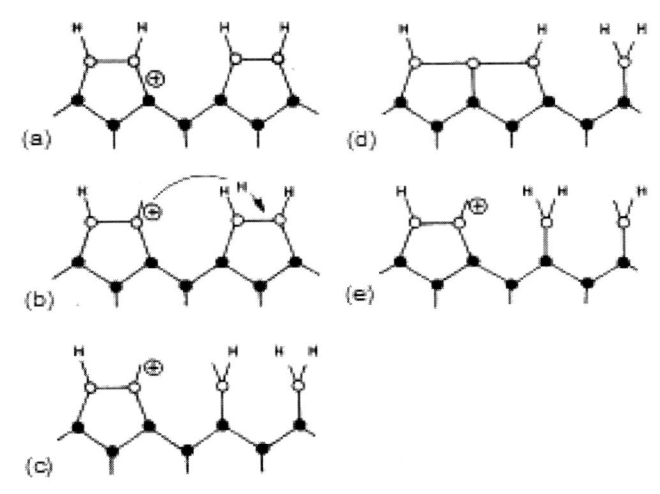

Figure 3.49   Schematic representation of the internal surface of a microvoid, where Si atoms at the surface are represented by open circles. [Reproduced from Carlson, *Appl. Phys. A*, **41**(1986) by permission of Springer.]

(4) Charged center trapping model

The positively charged threefold-coordinated silicon (T_3^+) and the negatively charged threefold-coordinated silicon (T_3^-) are candidates of charged trapping centers. Adler [1983] proposed that these charged centers capture electrons and holes, respectively, and, as a result, neutral silicon-dangling bonds are formed under illumination. It is supposed that these charged centers have negative-correlation energy, so-called negative-U centers. This has been derived from theoretical calculations by Bar–Yam and Joannopoulos [1986]. On the contrary, spatial potential fluctuation brings us to stabilize T_3^+ and T_3^- even if they have positive correlation energy.

(5) Impurity-involved model

Impurities such as nitrogen, oxygen, and carbon with more than a certain amount enhance the light-induced defect creation rate. A model [Redfield and Bube, 1990] involving nitrogen (threefold-coordinated dopant) is shown in Fig. 3.50(a,b). Ishii *et al.* [1985] have proposed a model involving an impurity atom such as oxygen being incorporated into a-Si:H as a positively charged threefold-coordinated oxygen, as shown in Fig. 3.50(c). Bandgap-illumination creates photo-electrons being captured by these centers that impurities has been excluded by an experiment for those high-quality a-Si:H samples containing oxygen of 2×10^{15} cm^{-3}, carbon of $7{-}10 \times 10^{15}$ cm^{-3}, and nitrogen of 5×10^{14} cm^{-3}, which exhibit

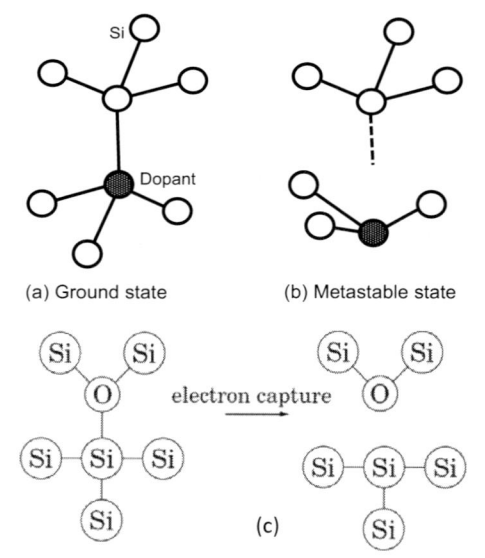

(a) Ground state   (b) Metastable state

electron capture

(c)

Figure 3.50 Atomic configuration of (a) fourfold-coordinated and (b) threefold-coordinated dopant (nitrogen). The dopant may be either a column-III acceptor or a column-V donor. The dashed line is a dangling bond . [Reproduced from Redfield and Bube (1990) by permission of Cambridge University Press.] (c) Schematic representations of atomic configurations: (Left side) Positively charged threefold-coordinated oxygen; (Right side) After electron capture, neutral twofold-coordinated oxygen and a Si-dangling bond are created [Reproduced from Shimizu, *Jpn. J. Appl. Phys.*, **43**, 3257 (2004) by permission of The Japan Society of Applied Physics.]

light-induced dangling bonds of as high as 5×10^{17} cm^{-3} [Kamei *et al.*, 1996].

3.5.3 Kinetics of Light-Induced Defect Creation in a-Si:H

3.5.3.1 Introduction

Stretched exponential relaxation is one of typical phenomena observed in disordered systems. The kinetics of light-induced defect creation observed after illumination is switched on, which is generally expressed by Eq. (3.62). The stretched exponential function is a solution of the following equation:

$$dN_d/dt = t^{\beta-1} (G - DN_d), \tag{3.66}$$

where G and D are the generation rate and a constant, respectively. In the following section, our model presented in Section 3.5.2 is found to give the kinetics of light-induced defect creation following the stretched exponential function.

Senda *et al.* [1999] attempted to use an alternative kinetic equation for the light-induced defect creation in amorphous semiconductors that is originally an expression of the logistic growth of population N in the biological system. It is given by

$$dN/dt = (k - \lambda N)N, \tag{3.67}$$

where k and λ are constants. This expresses a self-limiting process. Taking a saturated value of N as $N_S = N(\infty)$, Eq. (3.67) is rewritten as

$$dN/dt = k [1 - (N/N_S)]N, \tag{3.68}$$

where k corresponds to the time-dependent reaction rate, so that it is assumed to equate k with $At^{\alpha-1}$. Then, $N(t)$ is given by

$$N(t) = N(0)N_S \exp(t/\tau)^\alpha/\{N_S + N(0)[\exp(t/\tau)^\alpha - 1]\}, \tag{3.69}$$

Senda *et al.* [1999] applied the expression of Eq. (3.69) to illumination time-dependence of light-induced defects in a-Si:H and a-Ch.

3.5.3.2 Kinetics of light-induced defect creation in a-Si:H

First, we present calculated results, using Eqs. (3.36)–(3.49). Figures 3.51(a) and (b) are the calculated curves of r, q, and s as a function of illumination time for the case of a-Si:H no. 601211 with $G = 1.9 \times 10^{22}$ cm^3/s and a-Si:H no. 15y17 with $G = 4.0 \times 10^{22}$ cm^3/s, respectively. Figures 3.51(c) and (d) are the calculated curves of r, q, and s as a function of illumination time for the case of pm-Si:H no. 810091 with $G = 7.2 \times 10^{21}$ cm^3/s and pm-Si:H no. 601201 with $G = 2.7 \times 10^{22}$ cm^3/s, respectively. The values of parameters used in the calculation are listed in Table 3.1, in which the cases I, II, III, and IV correspond to (a), (b), (c), and (d) in Fig. 3.51, respectively. The $r + q$ vs. t curve is well fitted to a stretched exponential function given in Eq. (3.62), as shown in

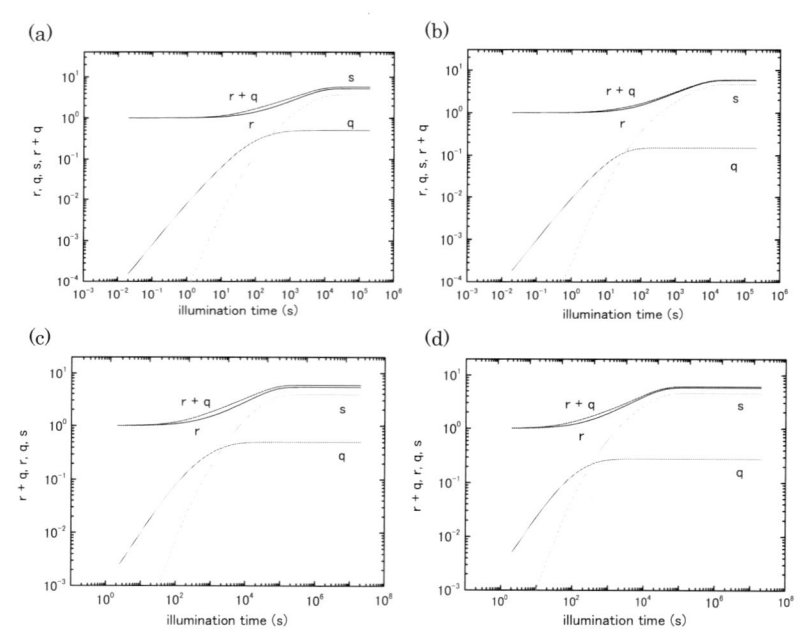

Figure 3.51 Densities of total dangling bonds, $r + q$, normal dangling bonds, r, hydrogen-related dangling bonds, q, and metastable hydrogen atoms, s, relative to N_{d0} as a function of illumination time under illumination with G. (a): a-Si:H ($T_s = 250°C$), No. 601211: $N_{d0} = 1.0 \times 10^{16}$ cm^{-3}, $G = 1.9 \times 10^{22}$ cm^{-3} s^{-1}, (b): a-Si:H ($T_s = 250°C$), No. 15y17, $N_{d0} = 1.0 \times 10^{16}$ cm^{-3}, $G = 4.0 \times 10^{22}$ cm^{-3} s^{-1}, (c): pm-Si:H ($T_s = 250°C$), No. 810091, $N_{d0} = 1.0 \times 10^{16}$ cm^{-3}, $G = 7.2 \times 10^{21}$ cm^{-3} s^{-1}, (d): pm-Si:H ($T_s = 200°C$), No. 601211, $N_{d0} = 1.8 \times 10^{16}$ cm^{-3}, $G = 2.7 \times 10^{22}$ cm^{-3} s^{-1}. For the values of parameters, see text. T_s: Substrate temperature.

Figs. 3.52(a)–(d). The values of β, τ, and N_{ss} are shown as a function of N_{ss} in Figs. 3.53(a) and (b) as well as the experimental results shown below, respectively. We note that the parameters β and τ characterize the features of the growing curve of N_d vs. t in the following way: β mainly determines the growing curve in the initial stage, that is, when β is small, N_d initially grows up rapidly, while τ determines the characteristic of the growing curve in the long term, that is, when τ is large, N_d grows up slowly in the long term. In our model, when G is large, β and τ become small, that is, N_d increases quickly at the initial stage of illumination and reaches the steady-state value, N_{SS}, for a short time.

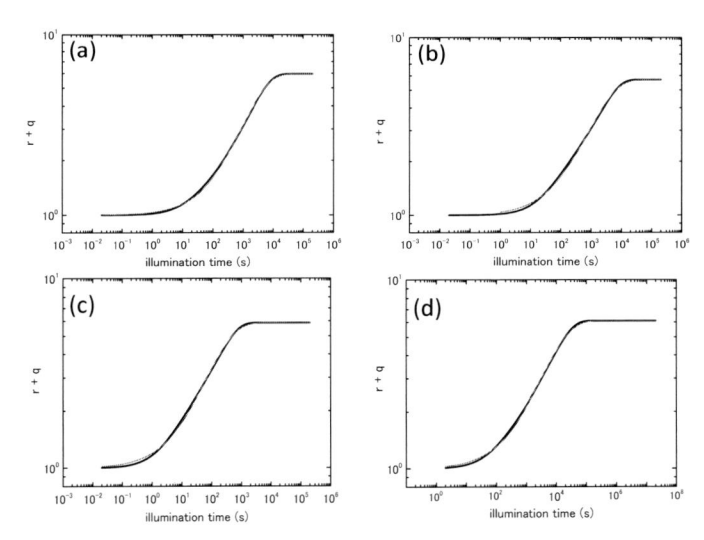

Figure 3.52 (a)–(d) Fitting of calculated curve (thick line) of $r + q$ vs. illumination time shown in Figs. 3.51(a)–(d) by a stretched exponential function (thin line) given by Eq. (3.62), respectively.

Table 3.1 The values of parameters used in the calculation.

Case	C_d (cm^3/s)	C_1 (cm^6/s)	C_2 (cm^3/s)	C_3 (cm^3/s)	C_4 (cm^3/s)	$N_{SS}\,(cm^{-3})$ $G \to 0$
I	2×10^{-15}	4×10^{-31}	1.28×10^{-21}	0	0	2.0×10^{16}
II	6×10^{-15}	4×10^{-31}	1.00×10^{-21}	0	0	1.3×10^{16}
III	2×10^{-15}	4×10^{-31}	1.43×10^{-22}	0	0	2.0×10^{16}
IV	2×10^{-15}	4×10^{-31}	1.43×10^{-22}	0	0	2.8×10^{16}
V	6×10^{-15}	4×10^{-31}	1.28×10^{-21}	0	0	4.8×10^{16}

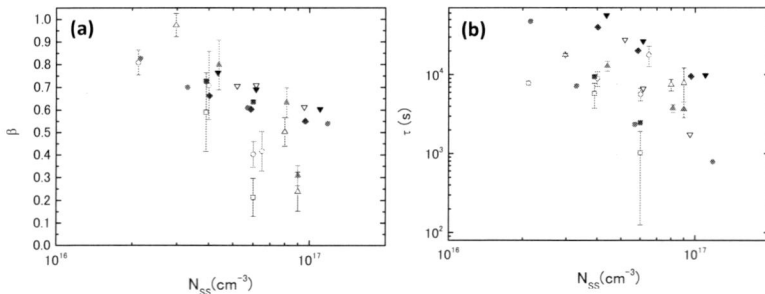

Figure 3.53 (a) Plots of the experimental and calculated values of β as a function of saturated dangling bond density N_{SS} in pm-Si:H and a-Si:H. Experimental results: Open triangles: pm-Si:H (T_s = 250°C), closed triangles: pm-Si:H (T_s = 200°C), open circles: a-Si:H (T_s = 250°C, Ecole Polytechnique), open squares: a-Si:H (T_s = 250°C, Yamaguchi University). Calculated results: Closed diamonds, obtained from the calculation (case III) for pm-Si:H (T_s = 250°C), closed inverse triangles, obtained from the calculation (case IV) and open inverse triangles, obtained from the calculation (case V) for pm-Si:H (T_s = 200°C). Calculated results: Closed circles, obtained from the calculation (case I) for a-Si:H (T_s = 250°C, Ecole Polytechnique), closed squares, obtained from the calculation (case II) for a-Si:H (T_s = 250°C, Yamaguchi University) [Morigaki *et al.*, 2010]. (b) Plots of the experimental and calculated values of τ as a function of saturated dangling bond density N_{SS} in pm-Si:H and a-Si:H. Experimental results: Open triangles: pm-Si:H (T_s = 250°C), closed triangles: pm-Si:H (T_s = 200°C), open circles: a-Si:H (T_s = 250°C, Ecole Polytechnique), open squares: a-Si:H (T_s = 250°C, Yamaguchi University). Calculated results: Closed diamonds, obtained from the calculation (case III) for pm-Si:H (T_s = 250°C), closed inverse triangles, obtained from the calculation (case IV) and open inverse triangles, obtained from the calculation (case V) for pm-Si:H (T_s = 200°C). Calculated results: Closed circles, obtained from the calculation (case I) for a-Si:H (T_s = 250°C, Ecole Polytechnique), closed squares, obtained from the calculation (case II) for a-Si:H (T_s = 250°C, Yamaguchi University) [Morigaki *et al.*, 2010].

For each illumination, the dangling bond density was measured as a function of illumination time and was fitted to the stretched exponential function given in Eq. (3.62). As examples, Figs. 3.41, 3.54(a) and (b) show the results for a-Si:H nos. 009151 and 601211 prepared at 250°C, pm-Si:H no. 810091 prepared at 250°C, and pm-Si:H no. 601201 prepared at 200°C, respectively. The values of β and τ obtained by such fittings are shown as a function of N_{ss} in a-Si:H and pm-Si:H in Figs. 3.53(a) and (b), respectively, along with those obtained from the calculation mentioned above. The value of N_{ss} depends on illumination intensity, that is, generation rate of free carriers, as shown in Figs. 3.55(a) and (b), in which the results obtained for a-Si:H and pm-Si:H are shown, respectively, along with the calculated curves. The calculated curves are obtained from Eqs. (24) and (25) of [Morigaki and Hikita, 2007] for the case of monomolecular recombination. The values of parameters used for the calculated curves I and II in Fig. 3.55(a) are cases I and II in Table 3.1, respectively, while those used for the calculated curves III–V in Fig. 3.55(b) are cases III–V in Table 3.1, respectively.

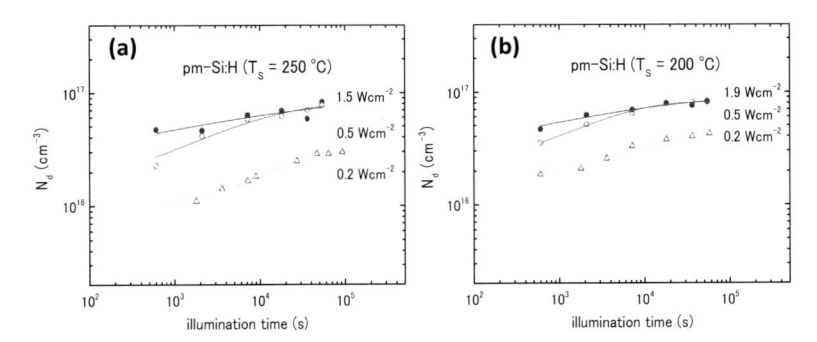

Figure 3.54 Plots of dangling bond density N_d *vs.* illumination time for various illumination intensities, (a) pm-Si:H (T_s = 250°C) No. 903181 for 0.2 W/cm², No. 810091 for 0.5 and 1.5 W/cm², (b) pm-Si:H (T_s = 200°C) No. 005092 for 0.2 W/cm², No. 601201 for 0.5 and 1.9 W/cm². The curve of N_d *vs.* illumination time is fitted by the stretched exponential function given by Eq. (3.62) with β and τ whose values are included in Figs. 3.53(a) and (b), respectively [Morigaki *et al.*, 2010].

From Figs. 3.51(a)–(d), it is found that the density of hydrogen-related dangling bonds is much lower than that of normal dangling bonds, similarly to the experimental results on the high-quality a-Si:H samples [Morigaki and Hikita, 2007; Morigaki *et al.*, 2007; Morigaki *et al.*, 2008].

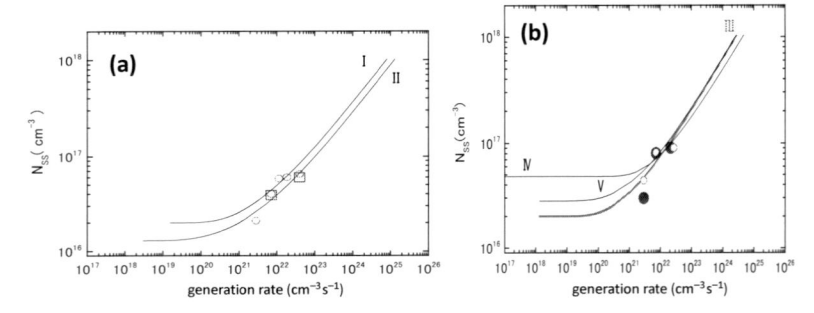

Figure 3.55 (a) Plots of saturated dangling bond density N_{SS} vs. generation rate of free carriers. (I) The calculated curve using the values of parameters in case I of Table 3.1. (II) The calculated curve using the values of parameters in case II of Table 3.1. Open circles: For samples prepared in Ecole Polytechnique, open squares: For samples prepared in Yamaguchi University. The values of N_{SS} in the zero-generation rate limit are 2×10^{16} cm^{-3} for curve a and 1.3×10^{16} cm^{-3} for curve β. The values of N_{SS} have an error of $\pm 25\%$ [Morigaki *et al.*, 2008]. (b) Plots of saturated dangling bond density *vs.* generation rate. Closed circles: pm-Si:H prepared at 250°C, open circles: pm-Si:H prepared at 200°C. The curves III–V are calculated ones. The curve III is a thin line, whereasile the curves IV and V are thick lines. For detail, see the text [Morigaki *et al.*, 2010].

In the following, we compare the calculated values of β and τ with their experimental values. For a-Si:H, the calculated results of β with the values of parameters for two cases I and II in Table 3.1 are similar to each other, while the experimental points for both types of samples have steeper decrease with increasing N_{SS} than the calculated points. The calculated results of τ with the values of parameters for cases I and II in Table 3.1 are similar to each other, while the experimental values for one type of samples (nos. 15y17, 15y18) have a tendency with N_{SS}, that is, decrease in τ with increasing N_{SS} is steeper than for calculated points and those for the other type of samples (nos. 009151, 601211) deviate from the calculated ones. For pm-Si:H prepared at 200°C, the calculated results of β with the values of parameters for two cases IV and V are similar to each other, whereas the experimental points are close to the calculated points except for that with high N_{SS}, that is, the experimental points have a steeper tendency with N_{SS} than the calculated points. For pm-Si:H prepared at 250°C, one can see a similar tendency to that for pm-Si:H prepared at 200°C. Concerning τ, the calculated points are close to the experimental ones except for low N_{SS}. For these comparisons, we note that the calculated results of β and τ can be obtained from the calculated curve of $r + q$ vs. t, which is derived from the rate equations with several parameters. The values of β and τ depend on the values of parameters, C_d, C_1, and C_2. Thus, the values of β and τ are determined in a complicated way, so that it is very difficult to obtain definitely good agreements between calculation and experiment.

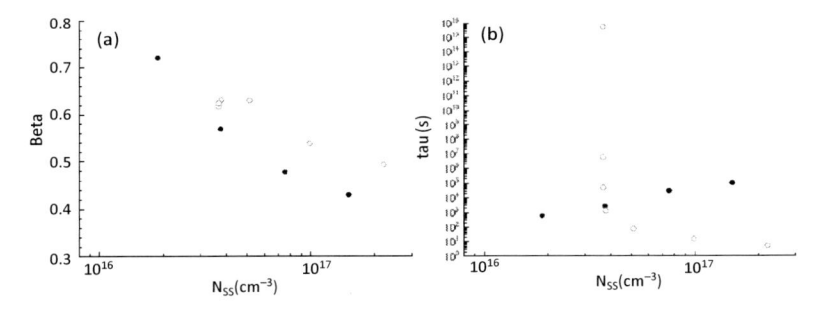

Figure 3.56 Calculated results of (a) β vs. N_{SS}, (b) τ vs. N_{SS}. Open circles and closed circles represent the calculated points by Morigaki and Hikita [2007] and Godet [1998], respectively [Morigaki and Hikita, 2007].

In the following, we discuss the behavior of β and τ with N_{SS} obtained from two models, that is, our model and the Godet model [Godet, 1998], in terms of calculated results, because the kinetics of light-induced defect creation has been shown to follow the stretched exponential relaxation by these two models, and a detailed comparison between calculation and experiment only by Godet model and our model is possible. Redfield [1989] pointed out only that experimental results obtained by other authors are fitted to the stretched exponential function. A comparison of calculated β and τ vs. N_{SS} between the above two models has been reported in [Morigaki and Hikita, 2007], as shown in Fig. 3.56. After then, we have obtained more experimental results in a-Si:H [Morigaki *et al.*, 2007, 2008] and furthermore in pm-Si:H

[Morigaki *et al.*, 2010]. Thus, it is worthy comparing further between two models here. The behavior of β and τ with N_{SS} in terms of our model has already been mentioned, although, in the Godet model, β becomes small, but τ becomes large with increasing N_{SS}, that is, N_d increases quickly at the initial stage of illumination, but, after then, it increases slowly to reach the steady-state value, N_{SS}, in comparison with our model [Morigaki and Hikita, 2007].

The above two models are different in the mechanism of the light-induced creation of metastable hydrogen atoms as follows: In the Godet model, metastable hydrogen atoms are generated directly by illumination from doubly hydrogenated (Si–H–H–Si) sites, while, in our model, they are generated from hydrogen-related dangling bonds, that is, in other words, they are generated indirectly by illumination through electron–hole recombination. Thus, in the Godet model, the termination of light-induced dangling bonds by metastable hydrogen atoms easily occurs compared with our model. The efficiency of the light-induced creation of dangling bonds is suppressed by such termination of light-induced dangling bonds by metastable hydrogen atoms (light-induced annealing) during illumination, so that τ in the Godet model may become long compared with our model. Thus, τ may be determined by competition between light-induced creation and annealing of dangling bonds. Light-induced annealing is more effective in the Godet model than in our model. The behavior of β with N_{SS} is similar for two models, but that of τ with N_{SS} is opposite, as was shown in [Morigaki and Hikita, 2007]. The experimental result on β agrees qualitatively with the calculated ones by two models. On the contrary, as shown in Fig. 3.53(b), τ decreases with increasing N_{SS} in a-Si:H nos. 15y17 and 15y18 and pm-Si:H, except for a-Si:H nos. 601211 and 009151 in which the values of τ are scattered and do not change so much with N_{SS}. From these observations, it may be concluded that our model seems reasonable to account for the behavior of β and τ with N_{SS} compared with the Godet model.

In the following, we discuss those differences in β and τ between a-Si:H and pm-Si:H that are correlated with the amorphous network. In the pm-Si:H film, dangling bonds exist in the amorphous network surrounding the nanocrystallite distributed with volume fraction of 2%. The amorphous network in pm-Si:H is more ordered than that in a-Si:H, so that this difference is expected to affect the values of β and τ. Actually, as shown in Figs. 3.53(a) and (b), the values of β for pm-Si:H are larger than those for a-Si:H, whereas the values of τ for pm-Si:H are similar to those for a-Si:H, taking into account that the experimental points obtained for a-Si:H are scattered. The value of β particularly determines the extent of distribution function of the lifetime of light-induced dangling bonds. The distribution function is the Fourier transform of the stretched exponential function (see the appendix). A higher value of β indicates a sharp lifetime distribution of light-induced dangling bonds with a well-defined lifetime close to the characteristic time τ. The fact that β becomes large compared with that for a-Si:H suggests that the lifetime distribution of normal dangling bonds becomes narrow compared with that for a-Si:H. This is simply understood by considering that the amorphous network of pm-Si:H is not complex compared with the amorphous network of a-Si:H. It has been suggested that the amorphous network of pm-Si:H is more ordered than that of a-Si:H [Roca i Cabarrocas *et al.*, 1998]. This is consistent with the above observation on the value of β.

From the above consideration, we note that β provides some indication concerning the amorphous network. The growing curve of light-induced dangling bond density with illumination time has been calculated for a-Si:H and pm-Si:H. For example, the calculated results corresponding to both materials prepared at 250°C have shown similar values of β for both materials at various illumination intensities. The calculated results based on the rate equations given by Eqs. (3.32)–(3.34) depend on the values of parameters C_d, C_1, and C_2, so that a unique determination of β is impossible, because the value of β is related to those of three parameters in a complicated way.

As shown in Fig. 3.53(a), β becomes small with increasing N_{SS}, that is, with increasing G in a-Si:H and pm-Si:H. We note that under strong illumination, β becomes small. This means that the lifetime distribution of normal dangling bonds becomes broad compared with under weak illumination or in the dark. This is connected to an increase in the dangling bond density. This is also related to the fact that the distance between a normal dangling bond and a hydrogen-related dangling bond becomes short with increasing dangling bond density, that is, annihilation rate of normal dangling bonds with metastable hydrogen atoms becomes high. In our model, the value of β can be determined by competition between light-induced creation of dangling bonds and annihilation of dangling bonds by metastable hydrogen atoms. The above tendency of β under illumination is consistent with the calculation. From the above consideration, it is known [Morigaki and Hikita, 2011] that the light-induced defect creation process is an example of stretched exponential relaxation processes in disordered system [Kakalios *et al.*, 1987].

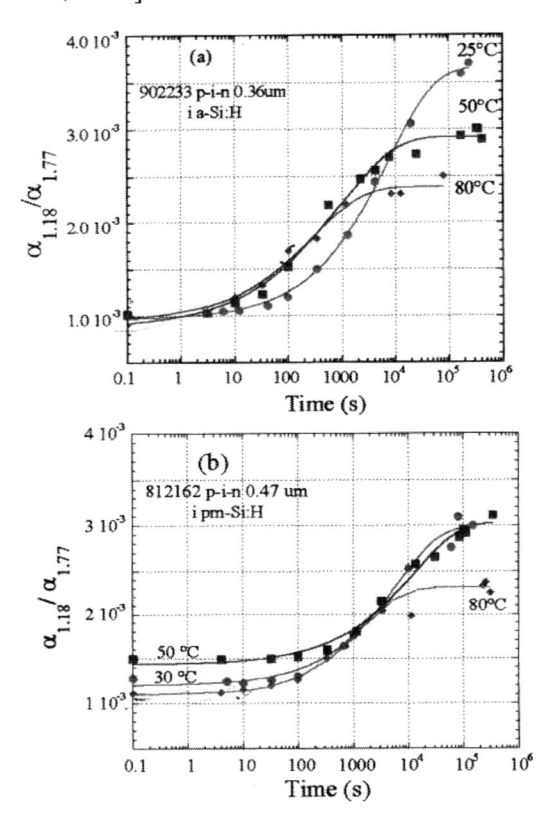

Figure 3.57 Plot of $\alpha_{1.18}/\alpha_{1.77}$ (the ratio of absorption coefficient at 1.18 eV to that at 1.77 eV) *vs.* illumination time (a) in i layer of a-Si:H of p-i-n solar cells and (b) in i layer of pm-Si:H of p-i-n solar cells. [Reproduced from Poissant *et al.*, Proceedings of the 16th European Photovoltaic and Solar Energy conference, 377 (2000) by permission of WIP.]

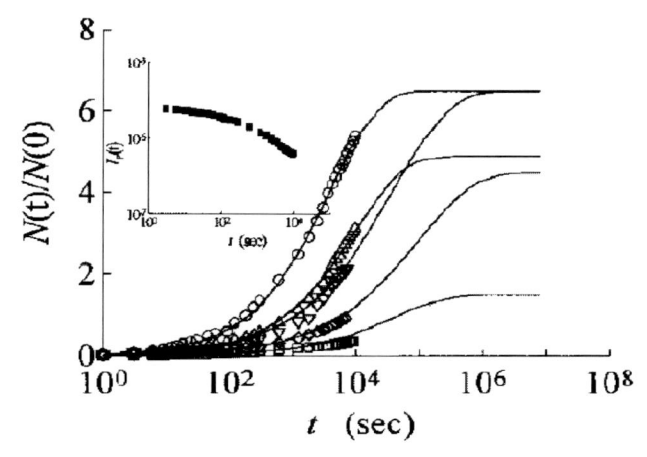

t (sec)

Figure 3.58 Plot of light-induced defect density relative to that at $t = 0$, $N(t)/N(0)$, *vs.* illumination time at 300 K in a-Si:H films prepared at 220°C by PECVD. Illumination at 2.41, 1.90, and 1.55 eV. Circle: 2.41 eV (5 mWcm^{-2}), triangle: 1.90 eV (30 mWcm^{-2}), inverse triangle: 2.41 eV (3 mWcm^{-2}), diamond: 1.80 eV (5 mWcm^{-2}), square: 1.55 eV (80 mWcm^{-2}). The solid curves are the calculated ones using a stretched exponential function. The inset shows an example of photocurrent *vs.* time for a photon energy of 1.90 eV. [Reproduced from Shimakawa *et al.*, *Philos. Mag. Lett.*, **84**, 81 (2004) by permission of Taylor & Francis.]

The dangling bond density can be normally measured by ESR, but it is estimated using CPM, photoconductivity, and modulated photocurrent method (MPC). From CPM measurements, change in optical absorption coefficient $\Delta\alpha$ for low-energy absorption (defect absorption) by illumination can be estimated, assuming that low-energy absorption is proportional to the dangling bond density N_d. In this case, N_d is overestimated because charged dangling bonds as well as neutral dangling bonds contribute to the low-energy absorption. For CPM measurements, two examples are shown in Figs. 3.57(a) and (b) [Poissant *et al.*, 2000], which are performed for the intrinsic layer of p-i-n solar cells having a-Si:H and pm-Si:H as the intrinsic layer, respectively. For p-i-n solar cells, the ratio of $\alpha_{1.18}/\alpha_{1.77}$ [absorption coefficient at 1.18 eV (defect absorption) and 1.77 eV (band-to-band absorption)] is taken. The curves of $\alpha_{1.18}/\alpha_{1.77}$ *vs.* illumination time are fitted to stretched exponential functions. The values of β and τ are shown in Figs. 3.60(a) and (b), respectively. For photoconductivity measurements, an example is shown in Fig. 3.58 [Shimakawa *et al.*, 2004], in which various photon energies of illumination are used. The curves of $N(t)$ *vs.* t are also fitted to stretched exponential functions. The estimated values of β and τ are shown in Figs. 3.60(a) and (b), respectively. These curves have been discussed in terms of our model [Morigaki *et al.*, 2008].

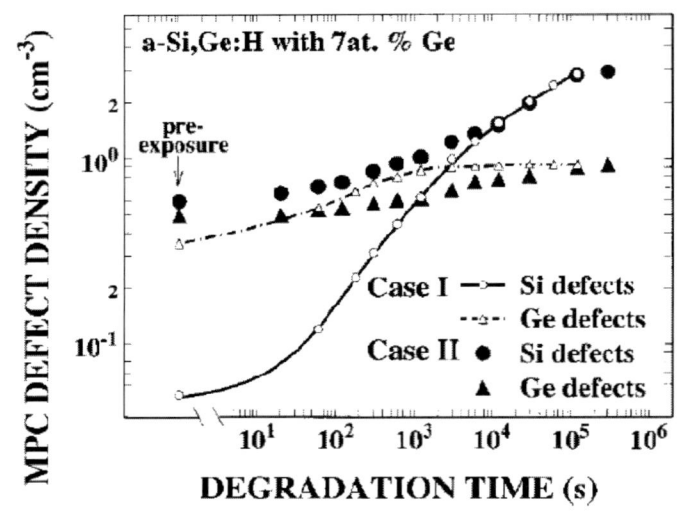

Figure 3.59　Plot of MPC defect density *vs.* degradation time in a-Si$_{0.93}$Ge$_{0.07}$:H films prepared by PECVD. Case I: Starting from the fully annealed state, Case II: Starting from the 420 K annealed state. Two types of defects start out nearly equal in densities. [Reproduced from Palinginis *et al.*, *Phys. Rev. B*, **63**, 201203-1 (2001) by permission of American Physical Society.]

Using MPC, densities of two types of dangling bonds, that is, Si-dangling bonds and Ge-dangling bonds, have been obtained in a-Si$_{1-x}$Ge$_x$ films with x = 2 and 7% by Palinginis *et al.* [2001], as shown in Fig. 3.59. One of their conclusions is that the result deduced from Fig. 3.59 is consistent with the bimolecular recombination mechanism by Stutzmann *et al.* [1985], but it is not consistent with the hydrogen-collision model [Branz, 1999].

The temperature variations of β and τ are shown in Figs. 3.60(a) and (b), respectively, which are collected from various literatures.

The temperature dependence of β associated with thermal annealing is also discussed in Section 3.5.9. The values of β and τ depend on temperature T in the following way [Kakalios *et al.*, 1987]:

$$\beta = T/T_0, \qquad\qquad (3.70)$$

$$\tau = \tau_0 \exp (E_a/k_B T), \qquad\qquad (3.71)$$

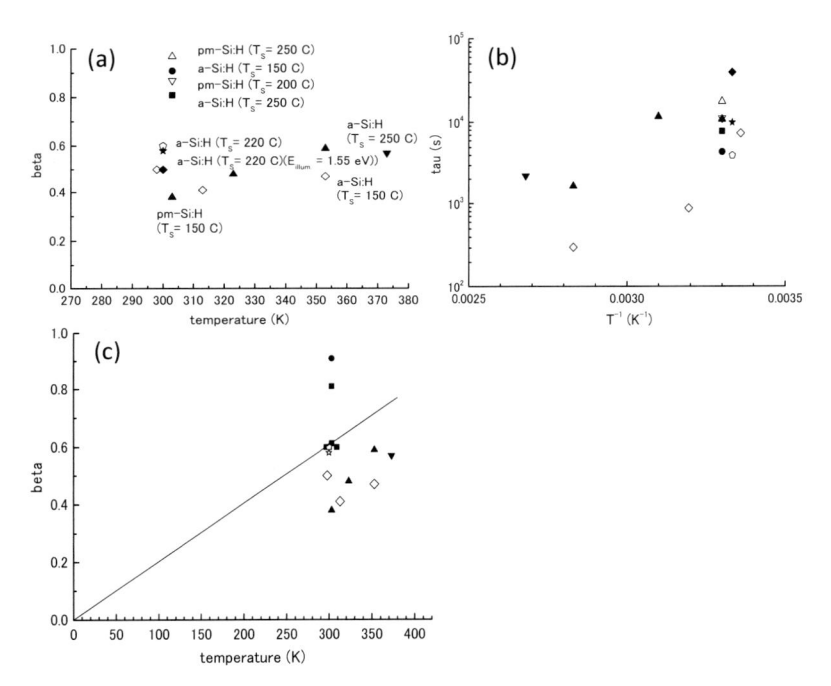

Figure 3.60 (a) Plot of β vs. T, (b) Plot of τ vs. T. The data are cited from the following literatures: Closed circle and closed square: [Morigaki *et al.*, 2007, 2008], open triangle and open inverse triangle: [Morigaki *et al.*, 2010], open pentagon, closed star, and closed diamond: [Shimakawa *et al.*, 2004], open diamond and closed triangle: [Poissant *et al.*, 2000], closed inverse triangle: [Godet, 1998]. (c) Plot of β vs. T. The solid line is the linear dependence of β on T as given by Eq. (3.70) with T_0 = 494 K.

Figure 3.60(c) shows a linear dependence on T, which is shown by a solid line with T_0 = 494 K. This value is compared with 600 K estimated in a-Si:H by Jackson and Kakalios [1988]. It should be noted that 600 K is estimated from annealing of light-induced defects along with the decay of excess carriers in n-type a-Si:H and the dispersive hydrogen diffusion in p-type and n-type a-Si:H, while 494 K is estimated from illumination-temperature dependence of β. However, the experimental points shown in Fig. 3.60(a) scatter, so that this dependence cannot be confirmed. The temperature dependence of τ almost follows Eq. (3.71), and thus, the estimated activation energy E_a is 192 meV that is compared with 0.94 eV in a-Si:H estimated by Jackson and Kakalios [1988]. However, these two values of E_a have different origins like the estimated value of T_0 in β, as mentioned above.

3.5.4 Light-Induced Hydrogen-related Dangling Bonds

Hydrogen-related dangling bonds have already been treated in Section 3.2.5. In this section, this issue is further dealt with from the viewpoint of light-induced defect creation. The presence of the hydrogen-related dangling bond, that is, a dangling bond having hydrogen in a nearby site, was suggested from a model of light-induced defect creation in a-Si:H [Morigaki, 1988]. Two types of dangling bonds, that is, the normal dangling bond and the H-related dangling bond, are created after illumination, as was mentioned before. It has been reported that the lineshape of the ESR signal due to dangling bonds does not change after prolonged illumination for high-quality samples. The reason for this observation is given as follows: The ESR signal is usually given by a derivative of the ESR line. The peak-to-peak separation of the ESR signal due to hydrogen-related dangling bonds is about twice larger than that due to normal dangling bonds, because the hyperfine dipolar interaction with hydrogen nucleus gives rise to the line-broadening associated with the hyperfine doublet. Thus, even if both types of dangling bonds have the same density, the peak-to-peak intensity due to the hydrogen-related dangling bond becomes approximately four times smaller than that due to the normal dangling bond. According to our calculation, the density of hydrogen-related dangling bonds created after illumination is one order of magnitude or several times lower than that due to normal dangling bonds, so that change in the lineshape of the ESR signal after illumination is either subtle or nothing. However, for low-quality samples containing a lot of hydrogen such as

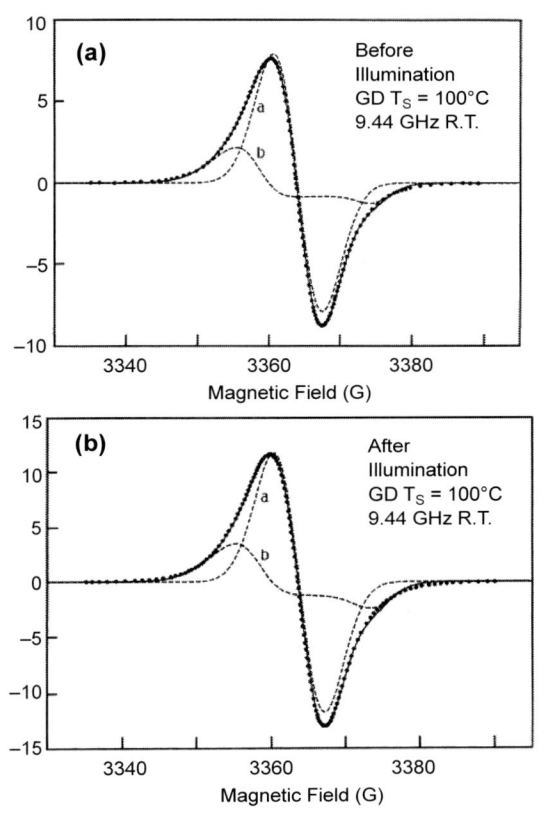

Figure 3.61 Deconvolution of the observed ESR spectrum for a-Si:H samples prepared at 100°C by PECVD. The solid points and dashed curves a and b are the observed spectrum and the calculated spectra due to normal dangling bonds and hydrogen-related dangling bonds, respectively. The solid curve is a superposition of dashed curves a and b. (a) Before illumination by an infrared light-cut xenon lamp of 600 mW/cm^2 for 6 h at room temperature. (b) After illumination by an infrared light-cut xenon lamp of 600 mW/cm^2 for 6 h at room temperature. The values of spin-Hamiltonian parameters used for fitting of ESR spectra are given in Table 3.2 (a-Si:H). Note that the same values are used for both cases before and after illumination. [Reproduced from Hikita *et al.*, *J. Phys. Soc. Japan*, **66**, 1730 (1997) by permission of The Physical Society of Japan.]

20–30 at.% or more, for example, for a sample prepared at 100°C ($N_{d0} = 4.9 \times 10^{17}$ cm^{-3}), the lineshape is deconvoluted into two components, as shown in Fig. 3.61(a). However, for low-quality samples containing a lot of hydrogen such as 20–30 at.% or more, for example, for a sample prepared at 100°C (N_{d0} = 4.9 × 10^{17} cm^{-3}), the lineshape is deconvoluted into two components, as shown in Figs. 3.61(a) and (b) before and after illumination for 6 h by a xenon lamp, respectively. In the sample, both types of dangling bonds are almost equally created after illumination and their density increases with a factor of 1.6 (N_d = 7.8 × 10^{17} cm^{-3}) compared with before illumination. Such a deconvolution of the ESR signal into two components has been done for low-quality a-Si:H and a-Si:D samples. For a-Si:H prepared at 100°C by PECVD of mixture gas of SiH_4 and H_2, the relative percentages of N_a and N_b are plotted as a function of gas-dilution ratio [SiH_4]/([SiH_4 + H_2]) in Fig. 3.21. These low-quality a-Si:H samples contain both types of dangling bonds almost equally. For a-Si:D, an example of deconvolution is shown in Fig. 3.62, in which two components are seen with the values of ESR parameters given in Table 3.2, taking into account the nuclear spin of ^{2}D = 1.

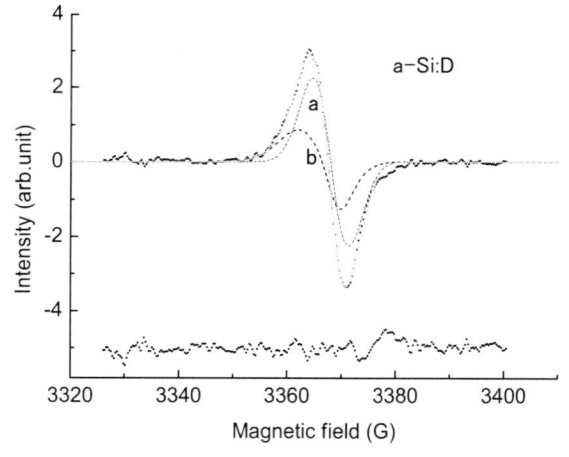

Figure 3.62 Deconvolution of the observed ESR spectrum for a-Si:D prepared by reactive sputtering with an rf power of 120 W in a mixture gas of Ar and D_2, whose ratio [D_2]/([Ar] + [D_2]) was 30%. The residual between the calculated curve and the observed spectrum is shown in the lower part of the figure with a scale twice extended compared with other curves. For the values of spin-Hamiltonian parameters, see Table 3.2 [Hikita *et al.*, unpublished].

Table 3.2   The values of parameters used in the calculation

	Normal DB (a)			H-related DB (b)							$N_s(\text{cm}^{-3})$
	g_a	$\sigma_a\,(G)$	$N_a(\%)$	$g_\parallel$	$g_\perp$	$A_\parallel(G)$	$A_\perp(G)$	$\sigma_\parallel(G)$	$\sigma_\perp(G)$	$N_b(\%)$	
a-Si:H	2.0051	3.5	60	2.0038	2.0068	15	2	2.0	5.3	40	4.9×10^{16}
a-Si:D	2.0039	3.4		2.0036	2.0060	1.55	0.3	1.2	4.5		

	$g_\parallel$	$g_\perp$	$\sigma_\parallel(G)$	$\sigma_\perp(G)$	$g_\parallel$	$g_\perp$	$A_\parallel(G)$	$A_\perp(G)$	$\sigma_\parallel(G)$	$\sigma_\perp(G)$
a-SiH (Astakov)	2.0043	2.0048	1.8	6.9	2.0021	2.0055	8.9	16.9	1.0	5.3

σ: Standard deviation of the Gaussian spin-packet.

The ENDOR measurements have been carried out for such low-quality a-Si:H and a-Si:D samples. An example of ENDOR spectra is shown in Fig. 3.63. For a-Si:H, two peaks at 2.7 and 10.7 MHz are seen in the figure, which are identified as being due to the distant ENDOR due to the ^{29}Si nucleus and the local ENDOR due to ^{1}H nucleus, respectively. For a-Si:D, a peak at 3.3 MHz is identified as being due to the local ENDOR due to the ^{2}D nucleus, because two frequencies of the local ENDOR are estimated from the following equation:

$$v_{ENDOR} = (1/2)\, a_D + v_N, \qquad\qquad (3.72)$$

where a_D and v_N designate the hyperfine interaction constant of the ^{2}D nucleus and the natural frequency of the ^{2}D nucleus, that is, 2.1 MHz, and a_D is estimated from a_H = 6 MHz to be 1.8 MHz. Then, the two ENDOR peaks are expected to be located at 1.64 and 2.56 MHz, so that the ENDOR peak at 3.3 MHz corresponds to the latter one and the former peak at 1.64 MHz seems to be overlapped with the distant ENDOR signal due to the ^{29}Si nucleus. For a-Si:H, the local ENDOR signal expected at 16.7 MHz is missing. This seems to be due to overlapping with the local ENDOR signal due to the ^{29}Si nucleus.

For electron-irradiated a-Si:H films, the hyperfine structure has been observed in the ESR spectrum of dangling bonds by Astakhov *et al.* [2008]. The spectrum is attempted to be deconvoluted into two components, as shown in Fig. 3.64. The values of ESR parameters estimated for this spectrum are listed in Table 3.2 as well as those obtained in other deconvolution.

A similar defect to the hydrogen-related dangling bond, that is, so-called vacancy–hydrogen complex, has been observed by Nielsen *et al.* [1997] in hydrogen-implanted c-Si, as mentioned before. A model of vacancy–hydrogen complex is illustrated in Fig. 3.22. Nielsen *et al.* [1997] reported that one of the principal axes, *z*, deviates from the direction [111] with 3°. Within the point-dipole approximation, they also estimated the distance between the H site and the dangling bond site to be 2.7 Å from an analysis of hyperfine interaction with H nucleus, comparing with the value of 2.8 Å calculated from the unrelaxed geometry of vacancy–hydrogen complex with a Si–H bond length of 1.5 Å.

Their values of hyperfine interaction constants are $A_x = -3.3$ MHz, $A_y = -4.6$ MHz, and $A_z = 8.5$ MHz, as shown in Table 3.2. In the case of a-Si:H, only the value of $A_\perp$ was estimated from the ENDOR measurement [Yokomichi and Morigaki, 1987; Yokomichi and Morigaki, 1996], that is, $|A_\perp| = 6$ MHz. This value is close to those of A_x and A_y described above.

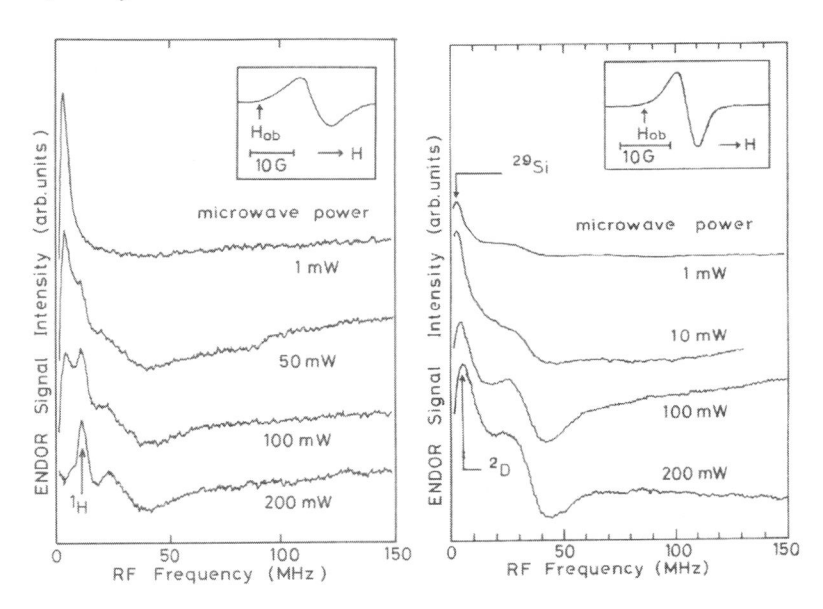

Figure 3.63 The left side: ENDOR spectra observed fo various microwave powers in a-Si:H samples prepared at 70°C by PECVD. The magnetic field is set at H_{ob} shown in the inset. [Reproduced from Yokomichi and Morigaki, *Solid State Commun.*, **63**, 629 (1987) by permission of Elsevier.] A local ENDOR signal due to ^{1}H nucleus is indicated by an arrow with ^{1}H. For low microwave power, a signal peak at 2.7 MHz corresponding to the natural nuclear frequency of ^{29}Si nucleus is a distant ENSOR signal due to ^{29}Si nucleus. The left side: ENDOR spectra observed for various microwave powers in a-Si:D samples prepared by reactive sputtering (see the figure caption of Fig. 3.62). The magnetic field is set at H_{ob} shown in the inset. A local ENDOR signal due to the ^{2}D nucleus and a distant ENDOR signal due to ^{29}Si nucleus are indicated by arrows. [Reproduced from Yokomichi and Morigaki, *Philos. Mag. Lett.*, **73**, 283 (1996) by permission of Taylor & Francis.]

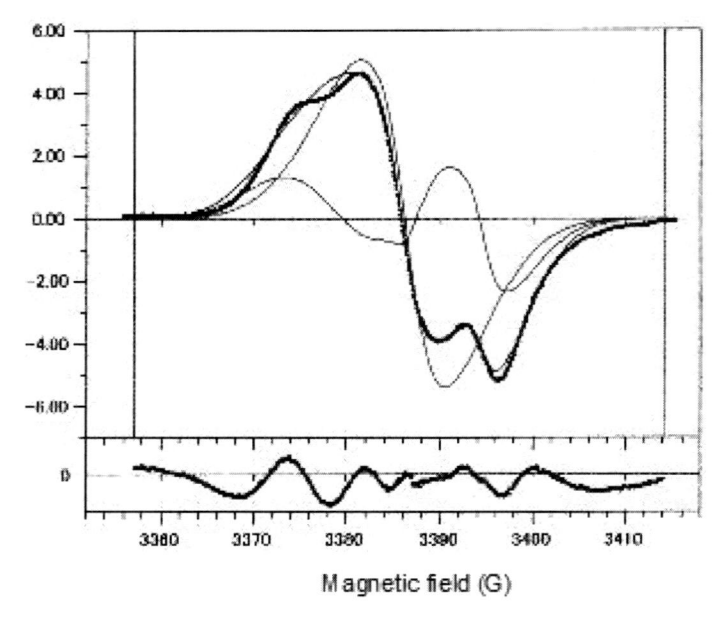

Magnetic field (G)

Figure 3.64 Deconvolution of the ESR spectra observed by Astakov *et al.*, [2008] into two components due to normal dangling bonds a and hydrogen-related dangling bonds b in an electron-irradiated a-Si:H sample. The solid points and the solid curve are the observed ESR spectrum and a superposition of dashed curves a and b, respectively. The residual between the calculated spectrum and the observed spectrum is shown in the lower part of the figure with a scale twice extended compared to other curves. [Reproduced from Morigaki *et al.*, *J. Optoelec. Adv. Mater.*, **11**, 1 (2009a) by permission of NIRDO.]

The distribution of hydrogen around the dangling bond has been investigated, using the pulsed ENDOR technique by Fehr *et al.* [2009]. Figure 3.65 shows their observation in which ENDOR efficiency is plotted as a function of $\nu_{RF} - \nu_H$, where ν_{RF} and ν_H are the radiofrequency used in the experiment and the natural frequency of hydrogen nucleus, respectively. They concluded that although the spectra indicate that hydrogen atoms can be located close ($r < 3$ Å) (r: the distance between the dangling bond and the hydrogen atom) to the dangling bond, no correlation between the dangling bond and the hydrogen distribution is observed and the hydrogen atoms are statistically distributed. However, we note that the spectrum has a shoulder at about −5 MHz, corresponding to a specific hydrogen atom close to the dangling bond. Such a spectrum can be constructed, using the computer

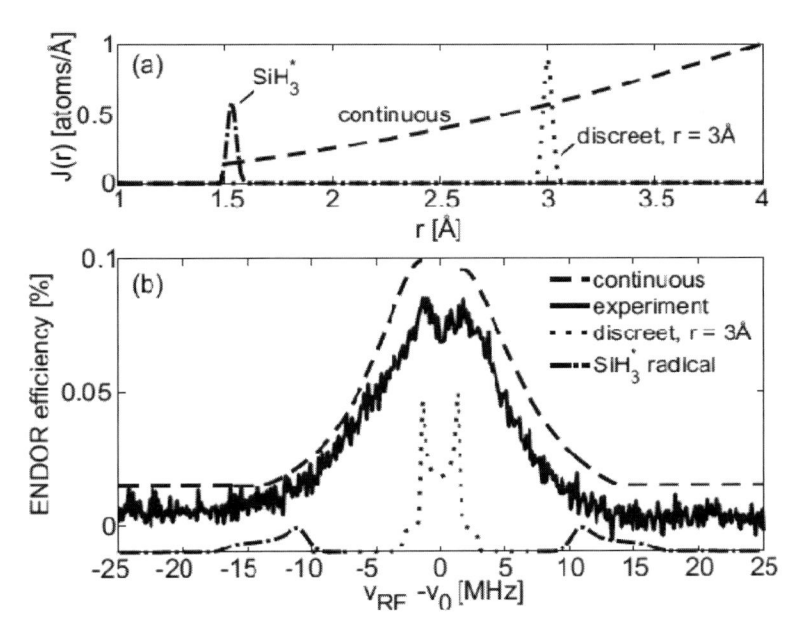

Figure 3.65 (a) Radial distribution function $J(r)$ for a continuous (dashed line) and discrete (dotted line) distribution of hydrogen atoms around dangling bonds. The position of hydrogen atoms in free silane radicals is shown by the dash-dotted line. (b) Corresponding ENDOR spectra for distributions in (a) and comparison with the experimental spectrum with an inversion pulse length of 40 ns. The dashed, the dotted and dash-dotted spectra are offset vertically with respect of the experimental spectrum for better overview. [Reproduced from Fehr *et al.*, *Phys. Status Solidi A*, **207**, 552 (2009) by permission of John Wiley and Sons.]

simulation as follows: From the analysis, we conclude that this specific hydrogen atom corresponds to the one adjacent to the dangling bond, that is, so-called hydrogen-related dangling bond, r, ranges between 2.1 Å and 2.7 Å from the analysis of deconvolution of the ESR line and also from the ENDOR experiment, that is, $r = 2.1$ Å, as mentioned before. Even for the hydrogen-related dangling bond, it interacts with hydrogen nuclei continuously distributed over the sample. It is worthy pointing out that the number of the hydrogen atom belonging to this dangling bond is only one, but continuously distributed hydrogen atoms exist statistically in proportion to r^2; it means that the number of distant hydrogen atoms increases with r, that is, the ENDOR signal intensity increases with r. This tendency easily hides the existence of the ENDOR signal due to the neighboring hydrogen atom in the hydrogen-related dangling bond. Furthermore, the density of hydrogen-related dangling bond is lower than that of normal dangling bond, particularly in high-quality a-Si:H films; the former is seven times or one order of magnitude lower than the latter. From the sample used by Fehr *et al.* [2009], the hydrogen-related dangling bond seems to exist relatively more than that for high-quality samples, because the sample by Fehr *et al.* was prepared at 160°C. Their result that hydrogen atoms exist close to the dangling bond, that is, $r < 3$ Å is remarkable, because until now, most of researchers put a constraint for constructing a model of light-induced defect creation in a-Si:H that there is no hydrogen atom within the distance of 3 Å from the dangling bond by citing the experiment by Isoya *et al.* [1993] and Yamasaki and Isoya [1992].

The distribution of hydrogen, for example, H–H distances, and hydrogen microstructure have been theoretically investigated by Timilsina and Biswas [2010].

3.5.5 Light-Induced Hydrogen Pairs

The NMR doublet due to hydrogen pairs has been observed only after optical excitation at 7 K by Su *et al.* [2002] in a-Si:H samples prepared at 200°C by PECVD, as shown in Fig. 3.66. The density of light-induced hydrogen pairs is between 10^{17} and 10^{18} cm^{-3}, which is consistent with the density of light-induced dangling bonds. This shows the direct evidence that hydrogen is involved in the light-induced creation of dangling bonds. This observation has been accounted for by Morigaki and Hikita [2005] on the basis of our model for light-induced defect creation in a-Si:H. In this model, hydrogen is dissociated from its original site nearby the hydrogen-related dangling bond under illumination. After then, this metastable hydrogen atom moves either to a normal dangling bond site to annihilate it or to another hydrogen-related dangling bond to terminate it for creating hydrogen pairs, as shown in Fig. 3.67. Such a process is expressed by the following rate equation,

$$dN_p/dt = C_4 N_m N_b, \tag{3.73}$$

where N_p designates the density of hydrogen pairs and the definitions of other parameters are the same as before [Eqs. (3.32)–(3.34)]. Combining with other rate equations given by Eqs. (3.37)–(3.39), we can solve numerically these rate equations with the following parameters: $r(t = 0) = 1$, $q(t = 0) = 0$, $A = 1.1 \times 10^{-2}$ s^{-1}, $A_1 = 6 \times 10^{-3}$ s^{-1} ($C_1 = 1.1 \times 10^{-31}$ cm^6/s). This is different from that reported in [Morigaki and Hikita, 2007] ($C_1 = 3 \times 10^{-31}$ cm^6/s) $A_2 = 1 \times 10^{-4}$ s^{-1}, $A_3 = 0$, $A_4 = 1 \times 10^{-2}$ s^{-1}, $A_5 = 0$ and $N_d(t = 0) = 1 \times 10^{16}$ cm^{-3}.

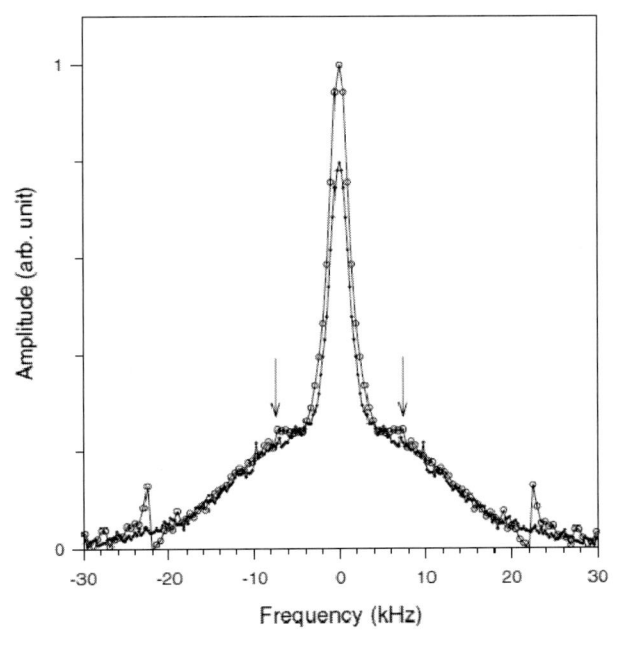

Figure 3.66 1H Jeener-Broekaert stimulated echo spectra in a-Si:H. Data for the irradiated and annealed (200°C for 4 h) samples are represented by open and solid circles, respectively. Arrows indicate the doublet in the irradiated sample. [Reproduced from Su *et al.*, *Phys. Rev. Lett.*, **89**, 015502-1 (2002) by permission of American Physical Society.]

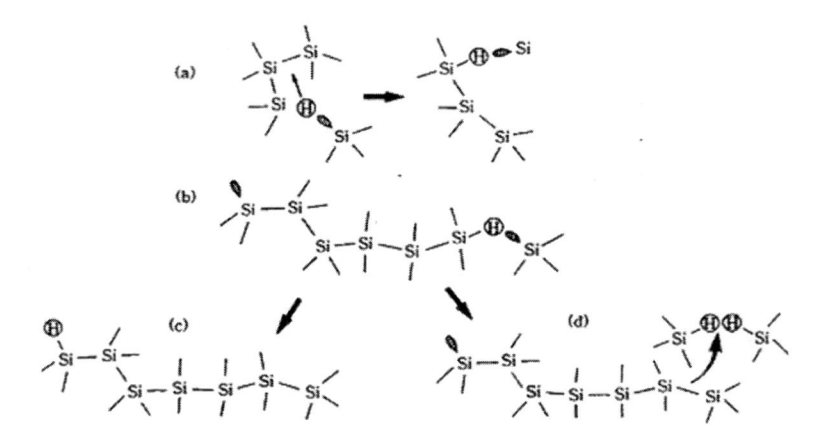

Figure 3.67 (a) Insertion of a hydrogen atom into a nearby weak Si–Si bond following the formation of a hydrogen-related dangling bond, (b) Dissociation of a hydrogen atom from a hydrogen-related dangling bond, (c) Termination of a normal dangling bond by a dissociated hydrogen atom, (d) Termination of a hydrogen-related dangling bond by a dissociated hydrogen atom and formation of a hydrogen-pair. [Reproduced from Morigaki and Hikita, *Solid State Commun.*, **136**, 616 (2005) by permission of Elsevier.]

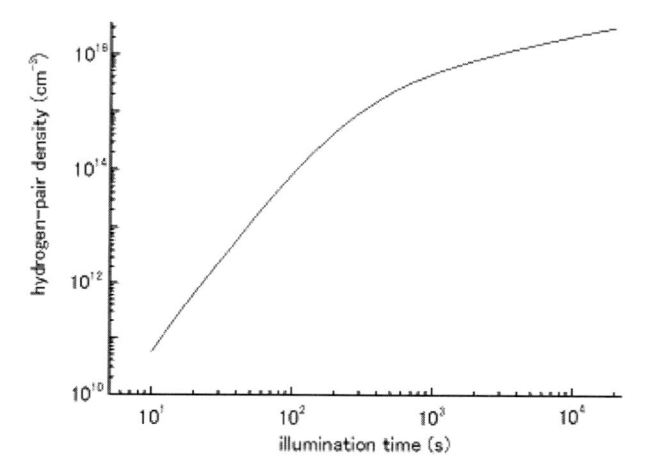

Figure 3.68 Calculated hydrogen-pair density as a function of illumination time, taking $N_{d0} = 1 \times 10^{16}$ cm^{-3}. For the values of parameters used in the calculation, see the text. [Reproduced from Morigaki and Hikita, *Solid State Commun.*, **136**, 616 (2005) by permission of Elsevier.]

The density of hydrogen pairs is shown as a function of illumination time t in Fig. 3.68. For comparison, r, q, s, and $r + q$ are also shown as a function of t in Fig. 3.69. From Figs. 3.68 and 3.69, the hydrogen pair is found to be created at a similar density to that of dangling bonds, being consistent with the observation of NMR. We note that $C_4 \neq 0$ is essential in this model to create the hydrogen pair, in which the saturation of light-induced dangling bonds does not occur, as shown before.

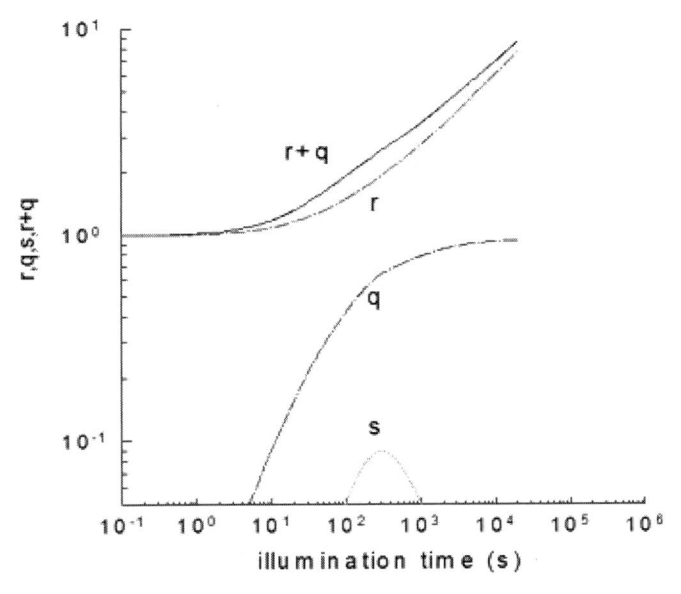

Figure 3.69 Calculated densities of total dangling bonds, $r + q$, normal dangling bonds, r, hydrogen-related dangling bonds, q, and metastable hydrogen atoms, s, relative to $N_{d0} = 1 \times 10^{16}$ cm^{-3} as a function of illumination time. For the values of parameters used in the calculation, see the text. [Reproduced from Morigaki and Hikita, *Solid State Commun.*, **136**, 616 (2005) by permission of Elsevier.]

Su *et al.* [2002] reported the spin-lattice relaxation time T_1 of hydrogen pairs as a function of the inverse of temperature in light-soaked a-Si:H, as shown in Fig. 3.70. The value of T_1 of hydrogen pairs increases rapidly with decreasing temperature from 7 K beyond the value of T_1 of bonded hydrogen, while T_1 of hydrogen pairs is shorter than that of bonded hydrogen above

6.5 K. T_1 of bonded hydrogen has been accounted for in terms of spin diffusion of the nuclear Zeeman energy to a molecular hydrogen relaxation center. Su *et al.* [2002] suggested that shortening of T_1 of hydrogen pairs above 6.5 K compared with bonded hydrogen may be due to the presence of the relaxation center such as Si-dangling bonds nearby the hydrogen pairs. In our model, such a Si-dangling bond corresponds to a normal dangling bond, being separated from the hydrogen pair site with a distance of about 13 Å as suggested from the mutual distance of a normal dangling bond and an H-related dangling bond. For bonded hydrogen, these normal dangling bonds may be effective as relaxation centers if they are located near bonded hydrogen, but, otherwise, they may be ineffective compared with ortho-hydrogen molecules.

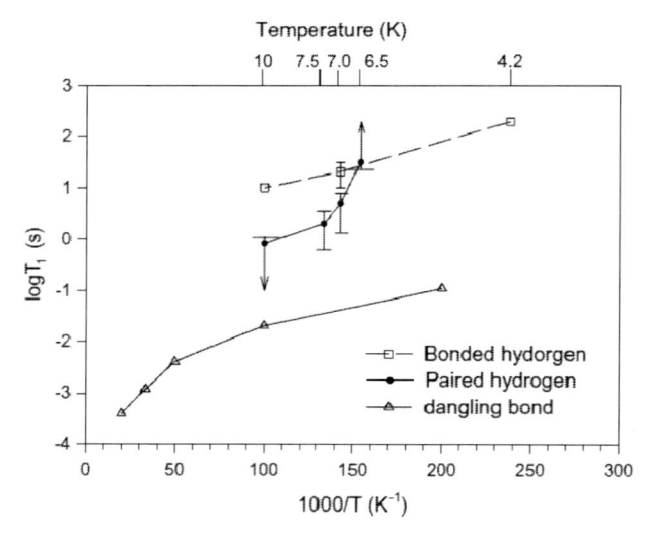

Figure 3.70 Temperature dependence of T_1 for the hydrogen doublet and the bonded hydrogen. The open squares represent the data for bonded hydrogen. [Stutzmann and Biegelsen, 1983] The solid circles represent the data for the hydrogen doublet. The open triangles are ESR data for the silicon dangling bond defect. The solid and dashed lines are aids to the eye. [Reproduced from Su *et al.*, *J. Non-Cryst. Solids*, **338–340**, 357 (2004) by permission of Elsevier.]

The geometric configuration of hydrogen pairs is also consistent with the Branz's collision model, but the problem of T_1 has not been discussed in terms of this model.

3.5.6 Low-Temperature Illumination

The model of light-induced defect creation postulates the diffusion of hydrogen atoms in the amorphous network. At low temperatures, such a diffusion should be slow, so that it affects the efficiency of defect creation under illumination. In this section, we discuss the case of low-temperature illumination. The observation of light-induced defect creation under illumination at low temperatures such as 2 K was first carried out by using the ODMR measurement [Morigaki *et al.*, 1982]. The light-induced dangling bond exhibits the quenching signal (as represented by D_2), as shown in Fig. 3.71. This signal is increased in intensity under illumination by argon-ion laser light at 514.5 nm (2.41 eV) and is decreased by raising temperature to 300 K, as shown in Fig. 3.71.

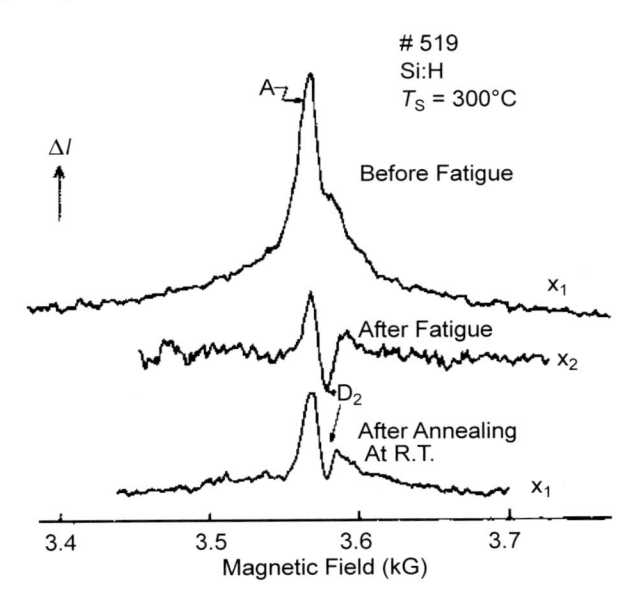

Figure 3.71 ODMR spectra taken at 2 K before and after fatigue (light soaking) and also after annealing at room temperature in an a-Si:H sample prepared at 300°C, monitoring the intensity of emitted light whose photon energy is 1.16 eV. A and D_2 represent the enhancing line due to the self-trapped holes and the quenching line due to the dangling bond, respectively. ΔI is the change in PL intensity associated with ESR. [Reproduced from Morigaki *et al.*, *J. Phys. Soc. Jpn.*, **51**, 147 (1982) by permission of The Physical Society of Japan.]

The efficiency of light-induced defect creation is actually low compared with that at room temperature. Two results of its illumination-temperature dependence are shown in Figs. 3.72(a) and (b), in which the former result and the latter one were obtained, using argon-ion laser light at 514.5 nm (2.41 eV) and ~1 W/cm^2, and 514.5 nm (2.41 eV) and 500 mW/cm^2, respectively. The latter result gives a value of the activation energy, nearly 10 meV, in the range of 65–340 K. The illumination-temperature dependence of the light-induced dangling bond density can be inferred from the measurement of the fill factor of a-Si:H solar cells. The fill factor is inversely correlated with dangling bond density. Figure 3.73 shows such results, that is, the fill factor normalized to that before illumination for various i-layers whose hydrogen concentrations are different ranging between 7.6 × 10^{21} and 1.3 × 10^{22} cm^{-3}. The illumination was carried out by an AM-1 light of an intensity of 500 mW/cm^2 with a duration of 5 h. As seen in the figure, the illumination-temperature dependence depends on the hydrogen content. The detail of the results is qualitatively understood in terms of our model of light-induced defect creation as follows: Under weak illumination, the light-induced dangling bond density at the saturation level is given by

$$\Delta N_{ds} = N_{ss} - N_{d0} \cong 2(C_d/C_1) \tag{3.74}$$

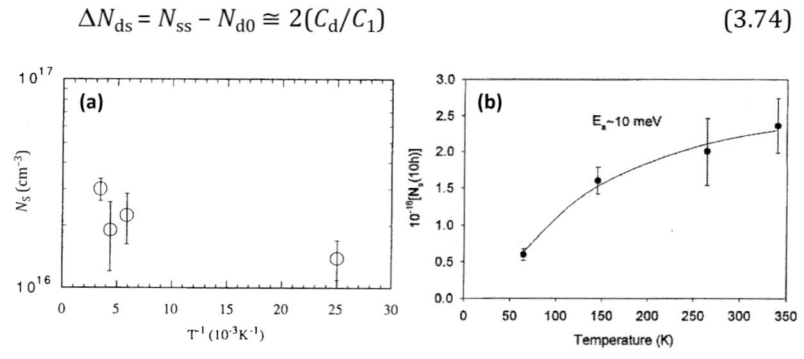

Figure 3.72 (a) Light-induced spin density as a function of the inverse of illumination temperature. [Replotted as N_S vs. $1/T$ from N_S vs. T plot of Fig. 1 in Yoshida and Taylor, 1992]. (b) Light-induced spin density $N_S(t)$ after 10 h of illumination at four temperatures. The activation energy of 10 meV is obtained from the curve of increase in $N_S(t)$. [Reproduced from Schultz and Taylor, *Mat. Res. Soc. Symp. Proc.*, **557**, 353 (1999) by permission of Materials Research Society.]

Namely, the light-induced defect creation competes the light-induced annealing of H-related dangling bonds by metastable hydrogen atoms. With raising temperature, this affects C_d and C_1, that is, this increases C_d and C_1. The illumination-temperature dependences of ΔN_{ds} depend on the detail of variations of C_d and C_1 with temperature (see Section 3.5.2).

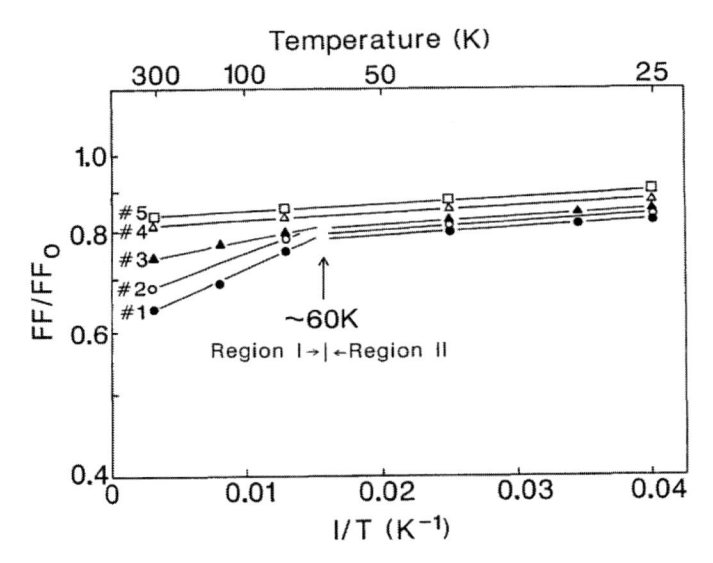

Figure 3.73 The fill factor of a-Si:H solar cells normalized to that before illumination as a function of the inverse of temperature. The hydrogen contents of the I layer are 26, 24, 20, 20, and 15 at.% for the cells #1, 2, 3, 4, and 5, respectively. [Reproduced from Nakamura *et al.*, *Jpn. J. Appl. Phys.*, **31**, 1267 (1992) by permission of The Japan Society of Applied Physics.]

Fuhs *et al.* [1985] measured illumination temperature-dependence of $\Delta L/L$ and ΔN_S, which are relative changes in PL intensity due to LS and light-induced dangling bond density, respectively, as shown in Fig. 3.74. The activation energy of ΔN_S is 4 meV. Stradins and Fritzsche [1996] measured illumination time-dependence of $\Delta \alpha/\alpha_0$ for subbandgap absorption relative to annealed state at 4.2 K for various illumination temperatures, using CPM with photon energy of 1.25–1.35 eV, as shown in Fig. 3.75.

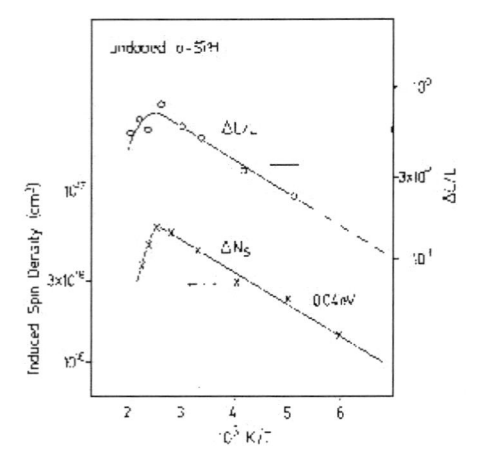

Figure 3.74 Light-induced spin density as a function of the inverse of illumination temperature. The relative change in the intensity of the defect luminescence, $\Delta L/L$, is shown for comparison. [Reproduced from Fuhs *et al.*, *Ann. Phys.*, **42**, 187 (1985) by permission of John Wiley and Sons.]

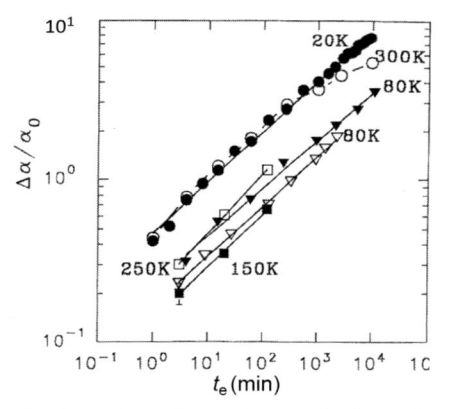

Figure 3.75 Illumination time dependence of the relative increase in subgap absorption relative to anneal state, $\Delta\alpha/\alpha_0$, measured at 4.2 K by CPM with photon energy of 1.25–1.35 eV due to light exposure of samples, GD (glow discharge) and HM (solid triangle), prepared by standard PECVD(GD) and by PECVD with a reactor having a heated mesh above the sample during deposition (HM), respectively, at different illumination-temperatures. The different symbols are used for different illumination temperatures. [Reproduced from Stradins and Fritzsche, *J. Non-Cryst. Solids*, **198–200**, 432 (1996) by permission of Elsevier.]

3.5.7 Pulsed Illumination

The measurements of light-induced defect creation in a-Si:H have been performed under intense illumination, using a pulsed laser light. The optical absorption and photoconductivity measurements have been done for investigations of light-induced defect creation under intense illumination. Two typical measurements are cited below. Tzanetakis *et al.* [1996] have observed the exposure-time dependence of $\Delta\alpha/\alpha_0$ for a-Si:H, using pulsed laser light of 5.2 mJ/cm^2/pulse (= 1.7×10^{16} cm^{-2}/pulse) at 1.91 eV and a repetition frequency of 7–10 Hz, as shown in Fig. 3.76, where $\Delta\alpha$ and α_0 are the change in absorption coefficient and the initial absorption coefficient, respectively. When the value of $PN^{1/2}$, in which P (= 52 mJ/cm^2/pulse) and N are the power per pulse and the number of pulse, respectively, is in the range of 0.1 and 1, the slope of the log $\Delta\alpha/\alpha_0$ vs. t, that is, $\Delta\alpha/\alpha_0 \propto t^\gamma$, is 0.67 for an overall least-square fit and 0.70 for a least-square fit with four experimental points, as shown in Fig. 3.76.

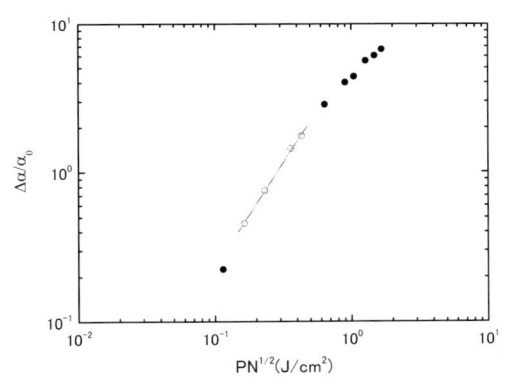

Figure 3.76 Relative change in subgap absorption $\Delta\alpha/\alpha_0$ as a function of the product of P (= 5.2 mJ/cm^2/pulse) and $N^{1/2}$, in which P and N are the power per pulse and the number of pulse, respectively (cited from [Tzanetakis *et al.*, 1996]). The solid line is a least-square fit of four experimental points (open circles) at small values except a smallest point. [Reproduced from Morigaki *et al.*, *J. Optoelec. Adv. Mater.*, **11**(1), 2009a, by permission of NIRDO.]

Stutzmann *et al.* [1994] have measured the illumination time-dependence of photoconductivity under pulsed illumination by a Nd-YAG pumped OPO and also by a Xe flash lamp, as shown in

Fig. 3.77(a) and (b), respectively. The photoconductivity, σ_p, may be anti-correlated with spin density, N_S, such as $\sigma_p \propto N_S^{-1}$ in the monomolecular recombination case. Actually, they measured a relationship between σ_p and N_S obtained from ESR measurements and found that this is the case for pulsed illumination by a Xe flash lamp for a 0.5 µm thick sample. The results of the above two measurements were compared with calculated results based on our model in good agreements [Morigaki *et al.*, 2009a].

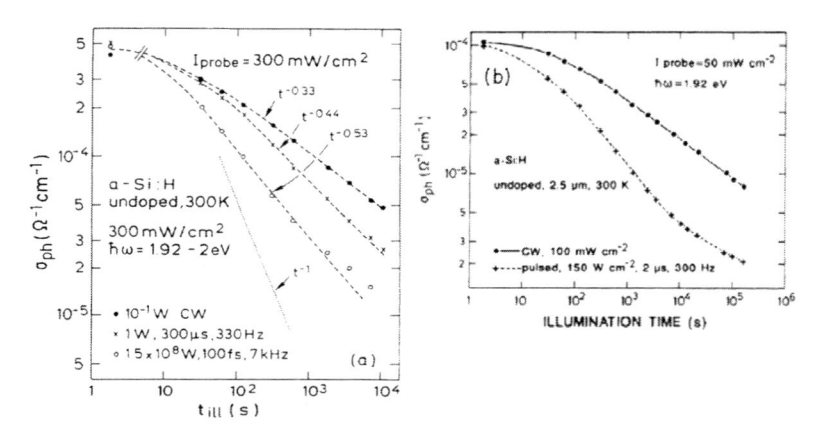

Figure 3.77 (a) Dependence of the photoconductivity on illumination time for the different (continuous (CW) or pulsed) light-soaking sources: (×) a chopped high intensity laser light with a pulse width of 300 µs, a repetition rate of 330 Hz, and a pulse power of 1 W/cm²: (○) a colliding pulse laser with a pulse width of 100 fs, a repetition rate of 7 kHz, and a peak pulse power of 0.4 GW/cm². The average light intensity was 300 mW/cm² and the incident photon energies were in a range 1.92–2.00 eV. (b) Comparison of photoconductivity fatigue curves for CW and pulsed illumination (Xe flash lamp) with white light of average intensity 100 mW/cm². The photoconductivity is monitored by defocused CW laser light (50 mW/cm², photon energy = 1.92 eV). [Reproduced from Stutzmann *et al.*, *Phys. Rev. B*, **50**, 11592 (1994) by permission of American Physical Society.]

Under intense pulsed illumination, bimolecular recombination plays an important role in determining the steady-state densities of electrons, n, and holes, p. Then, n and p are given by Eq. (3.50). Furthermore, we should take into account the number of weak Si–Si bonds, as shown in Eqs. (3.51)–(3.54).

Using these rate equations along with Eqs. (3.51)–(3.61), we calculate the densities of normal dangling bonds and hydrogen-related dangling bonds as well as the density of metastable hydrogen atoms. We compare the experimental results with the calculated ones [Morigaki *et al.*, 2003]. The ESR measurement was carried out at room temperature, while a pulsed illumination was performed at 10 K by using an YAG OPO laser system, in which pulsed light has a width of 10 ns, a repetition frequency of 11 Hz, and a power density of 100 mJ/cm^2 operating at either 2.48 (500 nm) or 1.55 eV (800 nm). The generation rates were 3.8×10^{30} and 1.4×10^{30} photons cm^{-3}s^{-1} for illumination at 2.48 eV and 2.0×10^{27} photons cm^{-3}/s for illumination at 1.55 eV. As mentioned in Section 3.5.2, in the calculation for pulsed illumination, the optical excitation of pulsed light is approximated by the continuous illumination, taking the total exposure time by pulsed light as the illumination time, because the rate equations involving pulsed optical excitation are not easy to be solved numerically. This is justified within an ambiguity of a factor of about 0.74 for the calculated light-induced defect density by citing the results of Stradins *et al.* [2000] in which they measured the defect creation efficiency for pulsed illumination as a function of the separation time of two laser pulses. Thus, numerical calculations are performed, using rate Eqs. (3.51)–(3.54), under continuous illumination. Figure 3.78 shows the illumination-time dependences of the relative densities of normal dangling bonds, r, H-related dangling bonds, q, metastable hydrogen atoms, s, and weak Si–Si bonds, w, under pulsed illumination at 1.55 eV, which are calculated using the following values of parameters: $G = 2.0 \times 10^{27}$ cm^{-3}s^{-1}, $C_d = 4 \times 10^{-14}$ cm^{3}s^{-1}, $C_1 = 1.1 \times 10^{-31}$ cm^{6}s^{-1},[1] $C_2 = 1 \times 10^{-20}$ cm^{-3}s^{-1}, $C_3 = 0$, $C_4 = 1 \times 10^{-18}$ cm^{3}s^{-1}, $C_5 = 0$, $N_{d0} = 1 \times 10^{16}$ cm^{-3} and $N_{w0} = 1 \times 10^{19}$ cm^{-3}. Here, we take into account that pulsed illumination corresponds to subgap illumination, so that the value of C_d is taken to be 10 times greater than that for bandgap illumination such as at 2.48 eV, because pulsed illumination at 1.55 eV directly creates a self-trapped hole in a weak Si–Si bond adjacent to a Si–H bond at low temperatures such as 10 K, taking into account the depth of self-trapped holes relative to the top of valence band, that is, 0.25 eV [Morigaki *et al.*, 1985] and the bandgap energy of 1.8 eV. The experimental points are also shown in the figure. For bandgap

---

[1]In Morigaki *et al.* (2003, 2009a), the value of C_1 is given as $C_1 = 1.1 \times 10^{-31}$ cm^{6}s^{-1}, as mentioned here.

illumination, the result is shown in Fig. 3.79, in which the values of parameters are the same as those in Fig. 3.78, except for C_d taken as 4×10^{-15} cm^{3}s^{-1} and G taken as 3.8×10^{30} photons cm^{-3}s^{-1}.

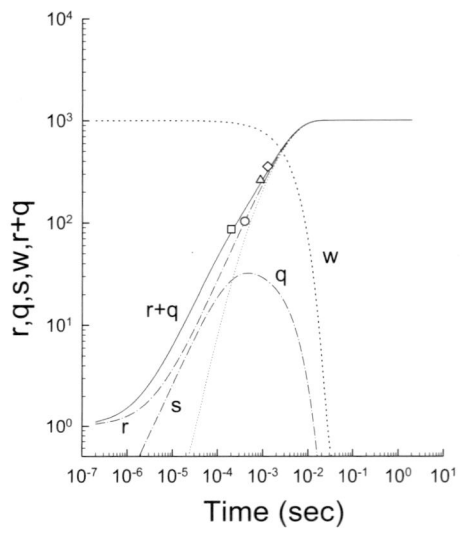

Figure 3.78 Densities of total dangling bonds, $r + q$ (solid line), normal dangling bonds, r (dot–dashed line), hydrogen-related dangling bonds, q (dot–dashed line), metastable hydrogen atoms, s (dotted line), and weak Si–Si bonds, w (dotted line), relative to $N_{d0} = 1 \times 10^{16}$ cm^{-3} as a function of total exposure time under illumination with $G = 2.0 \times 10^{27}$ cm^{-3}s^{-1} at 1.55 eV (800 nm): calculated curve and experimental points. For the values of parameters, see the text. The saturated dangling bond density of 1×10^{19} cm^{-3} is shown in the figure. [Reproduced from Morigaki *et al.*, *Philos. Mag. Lett.*, **83**, 341 (2003) by permission of Taylor & Francis.]

It may be interesting to see a difference in the lineshape between the ESR signals after pulsed and continuous illumination. We attempted to compare both spectra, as shown in Fig. 3.80, in which the left peak and right peak of the ESR absorption derivative curve were normalized in Figs. 3.80(a) and (b), respectively. As seen from these figures, difference in the observed lineshape between two types of illumination is obvious. The spectra taken after pulsed illumination were deconvoluted into two components, that is, normal dangling bonds and hydrogen-related dangling bonds, as shown in Fig. 3.81(a). The values of ESR parameters are given as follows: For normal dangling bonds, g = 2.0053, σ = 6.2 G, for hydrogen-related dangling bonds, $g_{||}$ = 2.0066, $g_{\perp}$ = 2.0090, $A_{||}$ = 12 G, $A_{\perp}$ = 3.2 G, $\sigma_{||}$ = 0.32 G, and $\sigma_{\perp}$ = 14.8 G. For 2 h of pulsed illumination under which the ESR line was observed, the exposure time is 5.04 × 10^{-6} s. For this time, we obtain r = 86.2, q = 46.6, and q/r = 0.54. The observed ratio of q/r from the deconvolution is 0.84, which is close to the calculated value of 0.54. The spectrum taken after continuous illumination is fitted by a component due to normal dangling bonds, as shown in Fig. 3.81(b), with the values of ESR parameters, $g_{||}$ = 2.0049, $g_{\perp}$ = 2.0062, $\sigma_{||}$ = 3.8 G, and $\sigma_{\perp}$ = 7.9 G. The result that only a component due to normal dangling bonds is observed in the ESR spectrum is consistent with the calculation, which shows that q is one order of magnitude smaller than r or several times smaller than r, depending on the values of parameters used in the calculation.

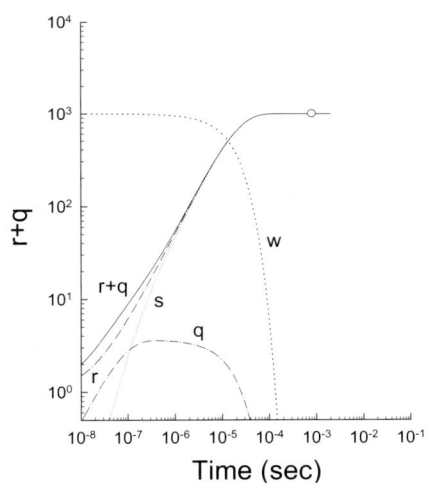

Figure 3.79 Densities of total dangling bonds, $r + q$ (solid line), normal dangling bonds, r (dot-dashed line), hydrogen-related dangling bonds, q (dot-dashed line), metastable hydrogen atoms, s (dotted line), and weak Si–Si bonds, w (dotted line), relative to N_{d0} = 1 × 10^{16} cm^{-3} as a function of total exposure time under illumination with G = 3.8 × 10^{30} cm^{-3} s^{-1} at 2.48 eV (500 nm): calculated curve and experimental points. For the values of parameters, see the text. The saturated dangling bond density of 1 × 10^{19} cm^{-3} is shown in the figure. [Reproduced from Morigaki *et al.*, *Phylos. Mag. Lett.*, **83**, 341 (2003) by permission of Taylor & Francis.]

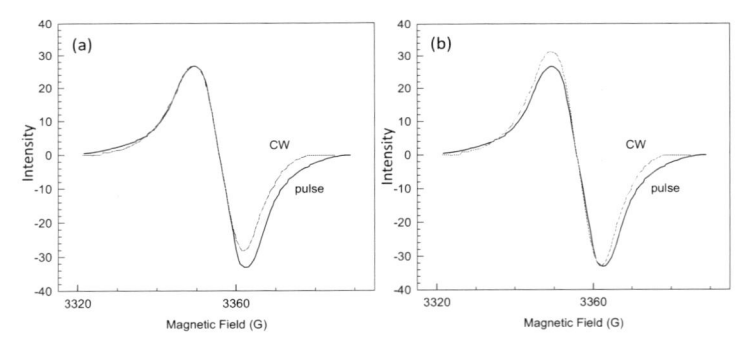

Figure 3.80 ESR spectra observed by Stutzmann *et al.* after 4 h of continuous (CW) illumination and after 2 h of illumination with a pulsed colliding pulse mode-locked laser (Pulse). [Reproduced from Stutzmann *et al.*, *Phys. Rev. B*, **50**, 11592 (1994) by permission of American Physical Society.] The average power and temperature are 300 W/cm^2 and 300 K, respectively. The low-magnetic field peak of ESR absorption derivative (a) and the high-magnetic field peak of ESR absorption derivative (b) are normalized for comparison of two spectra CW and pulse, respectively. [Reproduced from Morigaki *et al., J. Optoelec. Adv. Mater.* **11**, 1 (2009a) by permission of NIRDO.]

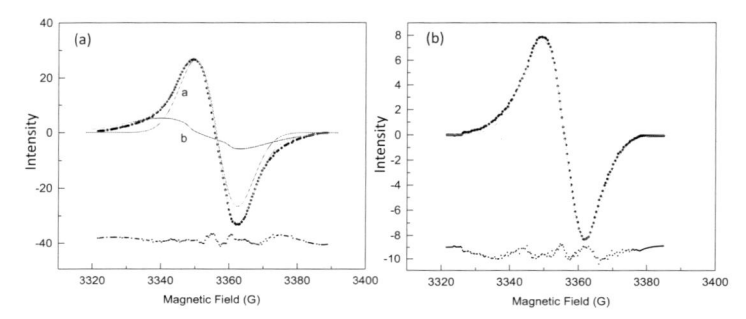

Figure 3.81 Deconvolution of the observed ESR spectrum into two components due to normal dangling bonds a and hydrogen-related dangling bonds b in a light-soaked a-Si:H sample with pulsed illumination shown in Fig. 3.80. The solid points and the solid curve are the observed ESR spectrum and a superposition of dashed curves a and b, respectively. The residual between the observed ESR spectrum and the calculated ESR spectrum is shown in the lower part of the figure with a scale twice extended compared with other curves. (a) Pulse in Fig. 3.80. (b) CW in Fig. 3.80: A component due to normal dangling bonds is taken into account. See the text for detail [Morigaki *et al.*, 2009a].

The ESR line corresponding to Figs. 3.78 and 3.79 is also deconvoluted into two components. It is found that the ESR spectrum is fitted by only one component due to normal dangling bonds with the values of ESR parameters: (a) $g_{||}$ = 2.0049, $g_{\perp}$ = 2.0050, $\sigma_{||}$ = 1.2 G, and $\sigma_{\perp}$ = 4.8 G, and (b) $g_{||}$ = 2.0050, $g_{\perp}$ = 2.0051, $\sigma_{||}$ = 1.2 G, and $\sigma_{\perp}$ = 5.3 G, as shown in Figs. 3.82(a) and (b). The result that the ESR spectrum is only due to normal dangling bonds is consistent with the calculated result, as shown in Figs. 3.78 and 3.79, that is, q is much smaller than r.

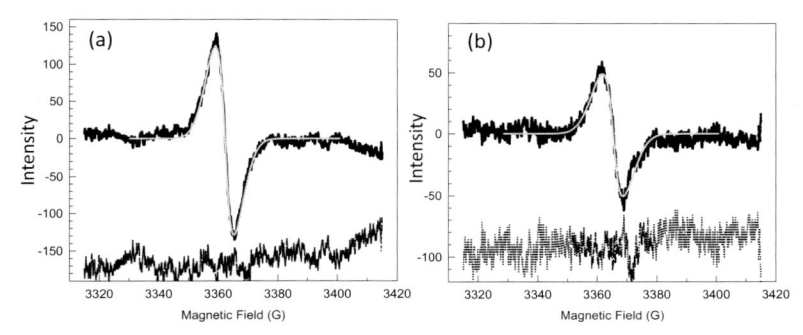

Figure 3.82 Fitting of the observed ESR spectrum by a component due to dangling bonds in a light-soaked a-Si:H sample with 1 h of pulsed illumination, (a) at 1.55 eV (800 nm) shown in Fig. 3.78, and (b) at 2.48 eV (500 nm) shown in Fig. 3.79. The solid points and the solid curve are the observed ESR spectrum and the calculated ESR spectrum, respectively. The residual between the observed ESR spectrum and the calculated ESR spectrum is shown in the lower part of the figure with a scale twice extended compared to other curves. A signal seen in the residual part in (b) is due to the E' center. [Reproduced from Morigaki *et al.*, *J. Optoelec. Adv. Mater.*, **11**, 1 (2009a) by permission of NIRDO.]

It is noted that the above high-quality sample exhibits an irreversible change in light-induced defect creation in contrast with the reversible change by continuous illumination, because thermal annealing up to 210°C could not anneal out light-induced dangling bonds, and thermally annealed dangling bond density saturated at 180°C, as shown in Fig. 3.83. Almost half of light-induced dangling bonds remain unannealed. This is the case of LS by sub-bandgap illumination at 1.55 eV, but a similar annealing behavior was observed for bandgap illumination at 2.48 eV. This irreversible process suggests that a strong local distortion occurs around the dangling bond, so that the dangling bond cannot be completely annealed out. For a low-quality a-Si:H sample prepared at 120°C containing dangling bonds of $5.3 \times 10^{18}\,\text{cm}^{-3}$ in the dark, intense pulse illumination causes a portion of native dangling bonds to be annealed even for thermal annealing at room temperature when the light-soaked sample is kept at room temperature for long time such as 1500 days (with a decay time of 250 days) after intense pulsed illumination at 10 K. This tells us a complicated effect of intense pulsed illumination on the sample, accompanying a local distortion of amorphous network [Morigaki and Ogihara, unpublished].

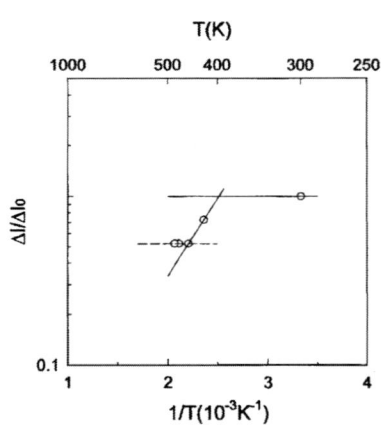

Figure 3.83 Light-induced dangling bond ESR intensity ΔI relative to the intensity ΔI_0 before thermal annealing *vs.* the inverse of annealing temperature in an a-Si:H film light soaked for 1 h illumination at 1.55 eV. The solid line and dashed line are the guides for the eyes. [Reproduced from Morigaki *et al.*, *Philos. Mag. Lett.*, **83**, 341 (2003) by permission of Taylor & Francis.]

3.5.8 Light-Induced Annealing of Dangling Bonds

The light-induced annealing of dangling bonds in a-Si:H is a sort of REDR being considered in Section 2.2. This has been first discussed by Redfield [1986] in the case of a-Si:H. Such a light-induced annealing effect has been observed by several authors, for example, see [Graeff *et al.*, 1993; Gleskova *et al.*, 1993; Zhang *et al.*, 1995; Vignoli *et al.*, 1996].

Here, we present the experimental evidence for light-induced annealing of dangling bonds taken by Takeda *et al.* [1997, 1998] in a-Si:H. Figure 3.84 shows three curves of the ESR peak-to-peak intensity as a function of illumination time for a-Si:H powder samples prepared at 70°C by PECVD, using pure silane gas, wherein continuous illumination was done at room temperature, using a high-pressure mercury lamp with an IR-cut filter at 1.3, 1.3, and 0.7 W/cm^2 for no. 2A, no. 2B, and no. 2C, respectively. As seen in the figure, the curve no. 2A clearly exhibits light-induced creation of dangling bonds at the initial stage, and then their light-induced annealing, while the other curves no. 2B and no. 2C exhibit only their light-induced creation.

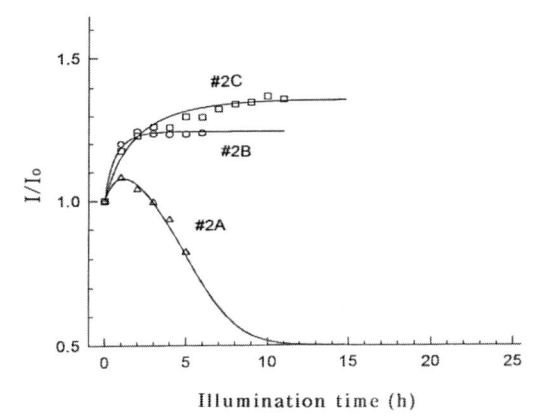

Illumination time (h)

Figure 3.84 Plot of the ESR intensity I relative to I_0 (I at $t = 0$) *vs.* illumination time in three a-Si:H samples #2A, #2B, and #2C. The solid curves are the calculated ones by Eqs. (3.75) and (3.76) (see the text). [Reproduced from Takeda *et al.*, *Jpn. J. Appl. Phys.*, **36**, 991 (1997) by permission of The Japan Society of Applied Physics.]

In order to interpret these results, we take the following model: The key process in the light-induced annealing is a nonradiative recombination of electrons with holes generated by illumination at hydrogen-related dangling bonds, which causes one or two hydrogen atoms to be dissociated from Si-H bond sites, as illustrated in Figs. 3.85(a)–(c). These hydrogen atoms diffuse over the amorphous network to terminate dangling bonds. The sample used in this study contains a number of dangling bonds, a number of hydrogen, and presumably a number of voids. The sample is divided into two regions, that is, the void region and the bulk region. In the void region, that is, around voids [called (a)], dangling bonds are not created under illumination, but annealed, while, in the bulk region, dangling bonds are created and annealed under illumination, [called (b)]. The configurations of dangling bonds in the void and bulk regon are shown schematically in Fig. 3.86, together with the schematic diagram of those kinetics. Thus, the rate equations are given for processes (a) and (b) in the following way: In the void region,

$$dN_{d1}/dt = -B_1 n p N_{d1}, \tag{3.75}$$

and, in the bulk region,

$$dN_{d2}/dt = (A_2 - B_2 N_{d2}) n p, \tag{3.76}$$

where N_{d1}, N_{d2}, n, and p designate densities of dangling bonds in the void region, dangling bonds in the bulk region, free electrons, and free holes including band-tail electrons and holes, respectively, B_1, B_2, and A_2 are the light-induced annealing coefficient in the void region and the bulk region, and the light-induced creation coefficient, respectively. For weak illumination, n and p are approximated by $G/\alpha N_d$, as given by Eq. (3.36), where G and N_d are the generation rate and the total dangling bond density $N_d = N_{d1} + N_{d2}$, respectively. The fitting values of parameters are given as follows: For no. 2A, $a_2 (= A_2 G^2/N_{d0}^3) = 0.4$ h⁻¹, $b_2 (= B_2 G^2/N_{d0}^2) = 0.8$ h⁻¹, and $b_1 (= B_1 G^2/N_{d0}^2) = 0.22$ h⁻¹. For no. 2B, $a_2 = 0.44$ h⁻¹, $b_2 = 1.8$ h⁻¹, and $b_1 = 0$. For no. 2C, $a_2 = 0.22$ h⁻¹, $b_2 = 0.58$ h⁻¹, and $b_1 = 0$, in which N_{d0} is N_d at $t = 0$. For nos. 2A and 2B, $G = 1.86 \times 10^{22}$ photons cm⁻³s⁻¹ $N_{d0} = 1.43 \times 10^{18}$ cm⁻³ and for no. 2C, $G = 1 \times 10^{22}$ photons cm⁻³s⁻¹, $N_{d0} = 1.43 \times 10^{18}$ cm⁻³. For nos. 2B and 2C, $b_1 = 0$ means that only the bulk region can contribute to light-induced effect. Under weak illumination (no. 2C), 0.7 W/cm², more dangling bonds seem to be created in the bulk region than in the void region. For no. 2B, the sample was obtained for the same source as the sample no. 2A, but the detail of voids (density, size, distribution, and so on) depends on sample location in the reaction chamber and other factors. Similar measurements are carried out for the film samples and the results were similar to those for the powder samples. The calculated N_d vs. t curves are shown in Fig. 3.84 in good agreements with the experimental curves.

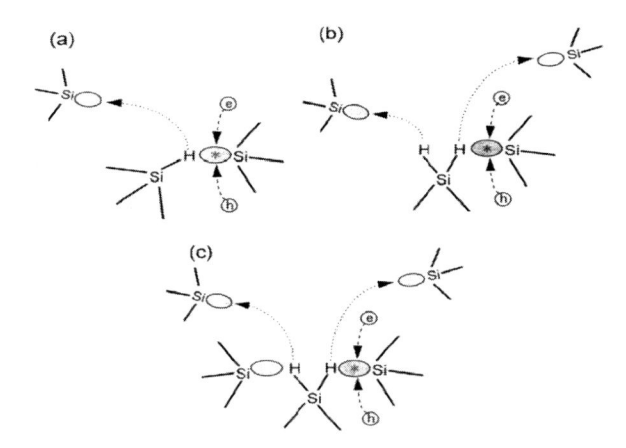

Figure 3.85 Schematic diagram illustrating (a) dissociation of one hydrogen atom from the Si–H bond and (b), (c) two hydrogen atoms from the Si–H_2 bond near the hydrogen-related dangling bond. After hydrogen dissociation, (a) one normal dangling bond and one hydrogen-related dangling bond are annihilated, (b) two normal dangling bonds and one hydrogen-related dangling bond are annihilated, and one normal dangling bond is ceated, and (c) two normal dangling bonds and two hydrogen-related dangling bonds are annihilated. [Reproduced from Takeda *et al.*, *Jpn. J. Appl. Phys.*, **36**, 991 (1997) by permission of The Japan Society of Applied Physics.]

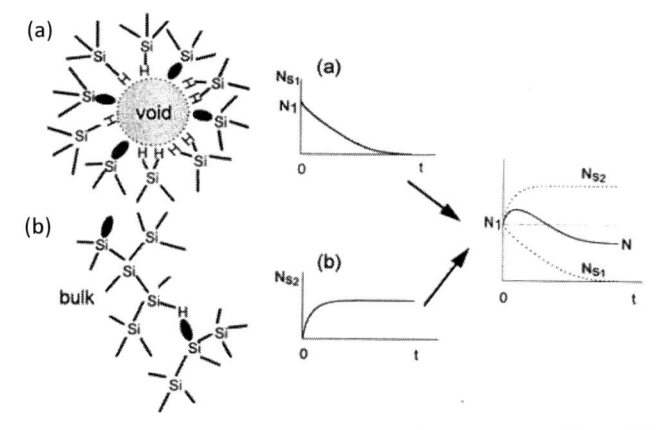

Figure 3.86 Schematic diagram illustrating the processes (a) and (b) and the illumination-time dependences of N_{s1} and N_{s2}. [Reproduced from Takeda *et al.*, *Jpn. J. Appl. Phys.*, **36**, 991 (1997) by permission of The Japan Society of Applied Physics.]

The ESR spectra for samples no. 2A and no. 2F were deconvoluted into two components, that is, normal dangling bonds and hydrogen-related dangling bonds, as shown in Figs. 3.87(a), (b) and 3.88(a), (b). Both samples were obtained from the same source, and are different in the illumination intensity I and duration t_d, that is, for 2A, $I = 1.3$ W/cm^2 and $t_d = 5$ h, and for 2F, $I = 0.7$ W/cm^2 and $t_d = 11$ h. From the deconvolution of ESR spectra taken before illumination, the normal dangling bonds constitute 61% of the total dangling bond density for both samples, while the hydrogen-related dangling bonds constitute 39% of the total dangling bond density for both samples. After illumination, the percentage density of normal dangling bonds and hydrogen-related dangling bonds relative to the total density depends on the illumination intensity and the illumination time (exposure time), that is, 58% and 42% for 2A [Fig. 3.87(b)], respectively, and 60% and 40% for 2F [Fig. 3.88(b)], respectively. This indicates that after weak illumination, the percentage density of both types of dangling bonds is almost the same, but after strong illumination, it becomes different, that is, hydrogen-related dangling bonds are more photocreated than

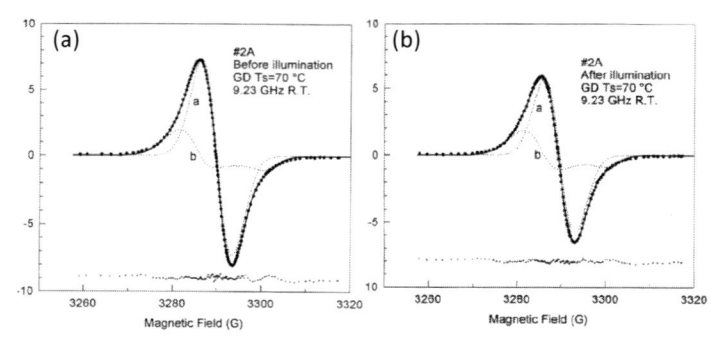

Figure 3.87 Deconvolution of the ESR spectrum observed (a) before illumination and (b) after illumination (1.3 W/cm^2, 5 h) into two components due to normal dangling bonds (dashed curve a) and hydrogen-related dangling bonds (dashed curve b) for sample #2A. The solid points and the solid curve are the observed ESR spectrum and the calculated ESR spectrum, respectively. The residual between the observed ESR spectrum and the calculated ESR spectrum is shown in the lower part of the figure with a scale twice extended compared to other curves. [Reproduced from Takeda *et al.*, *Jpn. J. Appl. Phys.*, **37**, 6309 (1998) by permission of The Japan Society of Applied Physics.]

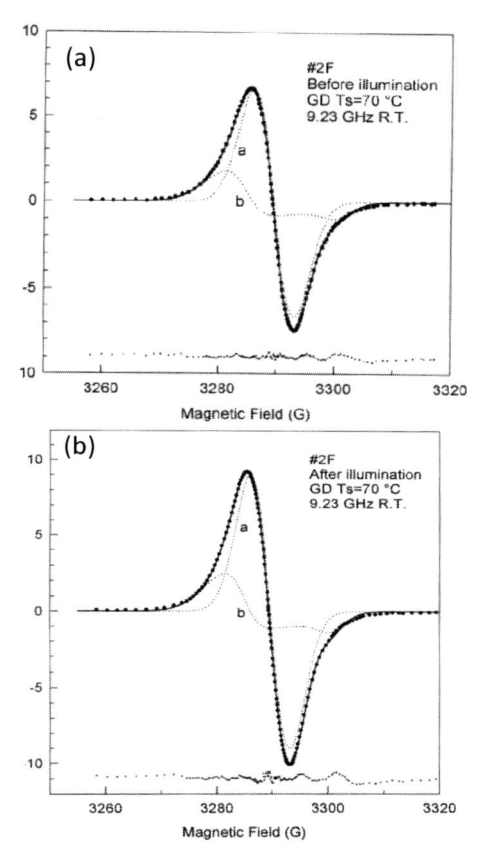

Figure 3.88 Deconvolution of the ESR spectrum observed (a) before illumination and (b) after illumination (0.7 W/cm^2, 11 h) into two components due to normal dangling bonds (dashed curve a) and hydrogen-related dangling bonds (dashed curve b) for sample #2F. The solid points and the solid curve are the observed ESR spectrum and the calculated ESR spectrum, respectively. The residual between the observed ESR spectrum and the calculated ESR spectrum is shown in the lower part of the figure with a scale twice extended compared to other curves. [Reproduced from Takeda *et al.*, *Jpn. J. Appl. Phys.*, **37**, 6309 (1998) by permission of The Japan Society of Applied Physics.]

normal dangling bonds. This result suggests that two hydrogen atoms are dissociated from the local configuration of the hydrogen-related dangling bond shown in Fig. 3.89(a), and then the previous hydrogen-related dangling bond is annihilated, but a new hydrogen-related dangling bond is created, as shown in Fig. 3.89(b). This results in that two normal dangling bonds are annihilated, but the number of hydrogen-related dangling bonds does not change, so that the percentage density of normal dangling bonds decreases, whereas that of hydrogen-related dangling bonds increases. This is consistent with the observation mentioned above. Thus, the model of existence of two kinds of dangling bonds, that is, those in the void region and in the bulk region is supported from the observation.

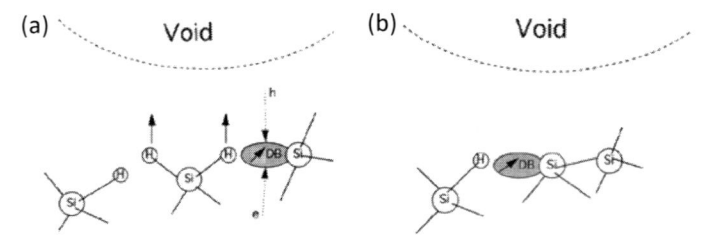

Figure 3.89   (a) Schematic diagram illustrating dissociation of two hydrogen atoms from the Si-H_2 bond near the hydrogen-related dangling bond as a result of nonradiative recombination of electrons and holes at the hydrogen-related dangling bond. (b) After hydrogen dissociation, two normal dangling bonds and one hydrogen-related dangling bond are annihilated, and a new hydrogen-related dangling bond is created as shown in the figure. [Reproduced from Takeda *et al.*, *Jpn. J. Appl. Phys.*, **37**, 6309 (1998) by permission of The Japan Society of Applied Physics.]

The distance between the nearby hydrogen and dangling bond site in hydrogen-related dangling bonds, R, can be deduced from anisotropic hyperfine interaction constant (see Section 3.2.5). The value of R does not change within the experimental inaccuracy of ±0.1 Å for both samples no. 2A and no. 2F between before and after illumination. This is consistent with the observation of light-induced structural changes in a-Si:H by Nonomura *et al.* [2000]. See also [Morigaki, 2001].

The values of ESR parameters are listed in Table 3.3 as well as the distance between nearby hydrogen and dangling bond site in hydrogen-related dangling bond, R, deduced from anisotropic hyperfine interaction constant.

Table 3.3   The values of parameters used in the calculation.

Sample No.	Normal DB(a)			H-related DB(b)						N_b (%)	N_s (cm^{-3})	R (Å)
	g_a	σ_a	N_a (%)	$g_{\parallel}$	$g_{\perp}$	$A_{\parallel}$	$A_{\perp}$	$\sigma_{\parallel}$	$\sigma_{\perp}$			
#2A[a]	2.0052	3.5	61	2.0035	2.0069	15.2	3.1	2.3	5.2	39	1.4×10^{18}	2.1
#2A[b]	2.0054	3.5	58	2.0034	2.0074	14.0	2.9	2.4	4.9	42	1.1×10^{18}	2.2
#2F[a]	2.0050	3.6	61	2.0034	2.0068	15.0	1.7	2.1	5.4	39	1.4×10^{18}	2.0
#2F[c]	2.0050	3.6	60	2.0030	2.0067	14.8	2.3	2.0	5.5	40	2.1×10^{18}	2.1

[a]Before illumination
[b]After illumination for 5 h at 1.3W/cm^2
[c]After illumination for 11 h at 0.7W/cm^2
*The unit of A and σ is gauss.

3.5.9   Thermal Annealing

The light-induced dangling bonds are thermally annealed in a-Si: H, in such a way that light-induced dangling bond density ΔN_d is decreased as a function of annealing time in the form of a stretched exponential function such as given below:

$$\Delta N_d(t) = \Delta N_d(0) \exp[-(t/\tau)^\beta]. \tag{3.77}$$

The values of β and τ depend on temperature T, following Eqs. (3.70) and (3.71). Jackson and Kakalios obtained $T_0 = 600$ K, and $\tau_0 = 1.8 \times 10^{-10}$ s and $E_a = 0.94$ eV for the light-induced dangling bond in high-quality a-Si:H films prepared at 230°C. The above value of T_0 was obtained from combined results with the decay of excess carriers in n-type a-Si:H and the dispersive hydrogen diffusion in p- and n-type a-Si:H. The relationship of Eqs. (3.70) and (3.77) is given by a model of a fractal-time random walk. During annealing, dissociated hydrogen is captured by the sites whose energy distribution is given by an exponential function of a characteristic temperature T_0, that is, the tail width is expressed by $k_B T_0$. The waiting time of hydrogen at a certain site is related to the hydrogen site energy distribution and, in a fractal-time random walk model, the waiting time distribution is related to β.

In the following, the values of β and τ for pm-Si:H are considered in comparison with those for a-Si:H [Morigaki *et al.*, 2005a, 2012; Morigaki and Hikita, 2010]. A powder sample of pm-Si:H, D no. 007171 was made of flakes collected from the reactor chamber after 16 h of deposition, which was done at 250°C by RF plasma decomposition of a mixture gas of SiH_4 (2%) and D_2 (98%). The spin density is 2×10^{16} cm^{-3}. The sample was illuminated at room temperature by a xenon-arc lamp of 500 W with an IR-cut filter at 1.6 W/cm^2. The observed decay curves are shown in Figs. 3.90(a) and (b), which are fitted by stretched exponential functions shown there. The obtained values of β and τ are shown as a function of annealing temperature and the inverse of annealing temperature in Figs. 3.91 and 3.92, respectively, as well as the values obtained by Jackson and Kakalios [1988]. In comparison with the values for a-Si:H, those for pm-Si:H are smaller for β and longer for τ than those for a-Si:H. For thermal annealing of native dangling bonds in pm-Si:H, their observed values of β and τ are shown in Figs. 3.93 and 3.94 as well as those for light-induced dangling bonds in pm-Si:H and in a-Si:H [Takeda *et al.*, 2008]. The case of native dangling bonds in pm-Si:H is also similar to that of light-induced dangling bonds in pm-Si:H compared with those in a-Si:H [Takeda *et al.*, 2000, 2008]. These results may be understood by taking into account the amorphous network of pm-Si:H. First, we note from the g-value consideration that dangling bonds in pm-Si:H exist in the amorphous network of pm-Si:H, although it contains nano-size crystallites. As mentioned before, the small value of β means that the distribution function of the inverse of lifetime is broadened, that is, the lifetime distribution for pm-Si:H is much broader than that for a-Si:H. The short-lifetime component corresponding to a rapid motion of metastable hydrogen atoms is involved in pm-Si:H. This is related to a relaxed amorphous network in pm-Si:H, as mentioned before. The long-lifetime component is also involved in pm-Si:H, so that it contributes to effective lifetime deduced from the stretched exponential function. That is related to the existence of hydrogen in the form of strongly bonded hydrogen molecules in platelets and voids. This is also related to light-induced defect creation in pm-Si:H during repeated cycles of illumination. The large value of T_0 for pm-Si:H compared with that for a-Si:H means that the hydrogen-site energy is distributed wider than that in a-Si:H. This is affected by the microstructure and the hydrogen-bonding scheme in amorphous network. For the slope of τ_0 vs. $1/T_a$, that is, E_a and τ_0, see [Morigaki *et al.*, 2005a].

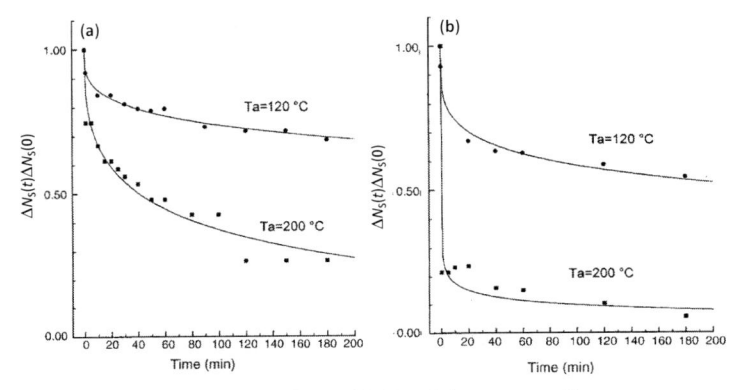

Figure 3.90 Decay curves of $\Delta N_s(t)/\Delta N_s(0)$ at annealing temperatures, 120°C and 200°C. (a) in air for sample no. 007171-2, (b) in vacuum for samples nos. 007171-5 and -6, respectively. The solid curves are stretched exponential functions given by Eq. (3.77), in which the values of β and τ are as follows: In (a), $\beta = 0.304$ and $\tau = 5.17 \times 10^3$ min for $T_a = 120$°C, and $\beta = 0.391$ and $\tau = 1.05 \times 10^2$ min for $T_a = 200$°C. In (b), $\beta = 0.263$ and $\tau = 1.08 \times 10^3$ min for $T_a = 120$°C, and $\beta = 0.127$ and $\tau = 1.33 \times 10^{-1}$ min for $T_a = 200$°C. [Reproduced from Morigaki *et al.*, *Mat. Sci. Eng. B*, **121**, 34 (2005a) by permission of Elsevier.]

For pm-Si:H prepared at 250°C, we have observed that the light-induced degradation efficiency decreases with repeated LS-100°C annealing cycles, while the dangling bond density reaches the same level after each annealing [Morigaki *et al.*, 2005a, b]. This is in a strong contrast with the case of a-Si:H [Hata *et al.*, 1998], in which the light-induced degradation efficiency measured from the sub-bandgap absorption coefficient after each LS-80°C annealing cycle in a-Si:H prepared at 250°C does not change, although the data were scattered, while sub-bandgap absorption coefficient after each annealing increased with repetition of the LS-80°C annealing cycle. The decrease in the light-induced degradation efficiency after the LS-annealing cycles is similar to those observed by Roy *et al.* [2002] for pm-Si:H prepared at 150°C (DOS measurements) and by Nickel *et al.* [1993] for polycrystalline Si:H (ESR measurements) (see Section 2.3), although the results in three types of experiments, that is, Roy *et al.* [2002], Nickel *et al.* [1993], and Morigaki *et al.* [2005a,b], for the annealing behavior in each LS-annealing cycle are different. We have concluded that the above result on the decrease in the light-induced degradation efficiency by repeating LS-100°C annealing cycles is attributed to an increase in the dangling bond-termination rate by mobile hydrogen with LS-annealing cycling, because hydrogen moves faster and faster, as a result of repeated cycles. Such a fast hydrogen movement is suggested from the observation that β decreases with repeated LS-annealing cycling, that is, the lifetime distribution of light-induced dangling bonds is broadened. The existence of light-induced dangling bonds with a short lifetime is related to the fast diffusion of hydrogen enhanced by ordering of the amorphous network, as mentioned before. The existence of strongly bound hydrogen proposed by Nickel *et al.* [1993] and Roy *et al.* [2002] is also suggested from the observation of light-induced dangling bonds with a long lifetime.

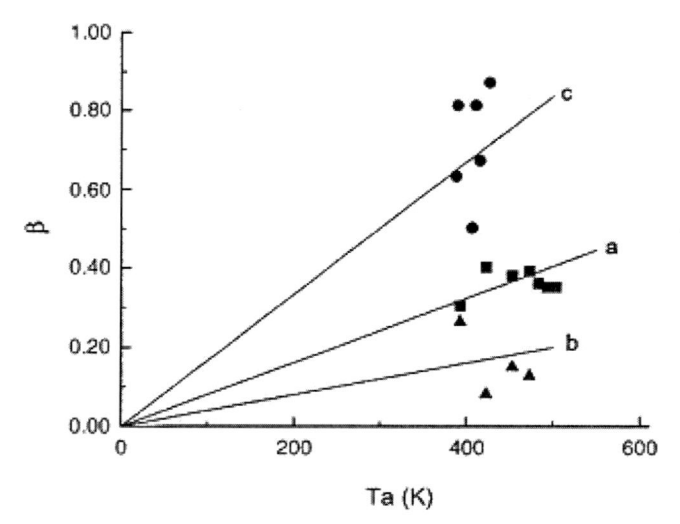

Figure 3.91 Plots of β *vs.* annealing temperature T_a; solid squares: Sample No. 007171 for annealing in air, solid triangles: Sample No. 007171 for annealing in vacuum and solid circles: Data for a-Si:H taken from [Jackson and Kakalios, 1988]. Three solid lines a, b, and c are the least-square fittings given by Eq. (3.70) for three cases, that is, annealing in air and in vacuum, and Jackson and Kakalios [1988], respectively. For detail, see the text [Morigaki *et al.*, 2005a].

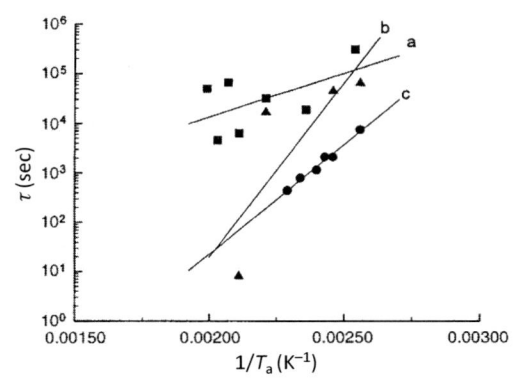

Figure 3.92 Plots of τ vs. annealing temperature T_a; solid squares: Sample no. 007171 for annealing in air, solid triangles: Sample 007171 for annealing in vacuum and solid circles. Data for a-Si:H taken from Jackson and Kakalios (1988). Three solid lines a, b, and c are the least-square fittings given by Eq. (3.71) for three cases, that is, annealing in air and in vacuum, and Jackson and Kakalios [1988], respectively. For detail, see the text. [Reproduced from Morigaki *et al.*, *Mater. Sci. Eng. B*, **121**, 34 (2005a) by permission of Elsevier.]

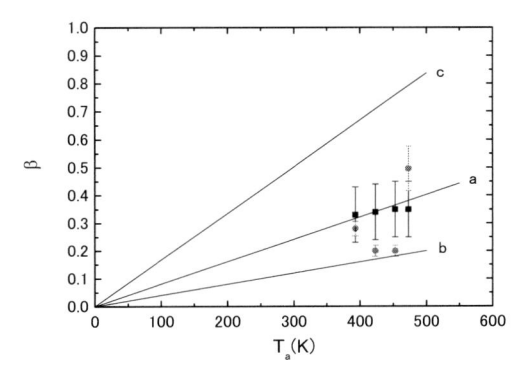

Figure 3.93 Plots of β vs. annealing temperature T_a. Closed squares: The estimated values for thermal annealing in air and closed circles: The estimated values for thermal annealing in vacuum. Three solid lines a, b, and c are the least-square fittings of the experimental points for three cases reproduced from Fig. 3.6 in [Morigaki *et al.*, 2005a], that is, thermal annealing in air and in vacuum of light-induced dangling bonds in pm-Si:H by Morigaki *et al.* [2005a] and thermal annealing of light-induced dangling bonds in a-Si:H by Jackson and Kakalios [1988], respectively [Takeda *et al.*, 2008].

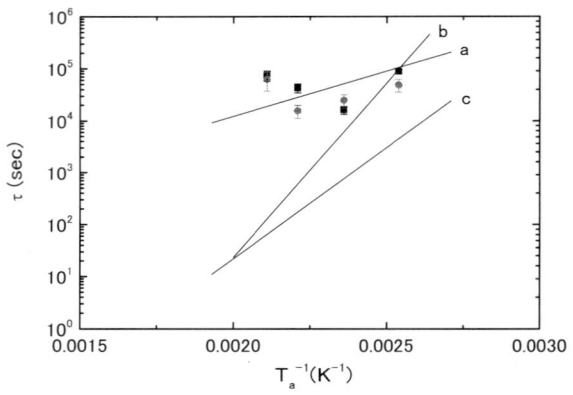

Figure 3.94 Plots of τ *vs.* the inverse of annealing temperature T_a. Closed squares: The estimated values for thermal annealing in air and closed circles: The estimated values for thermal annealing in vacuum. Three solid lines a, b, and c are the least-square fittings of the experimental points for three cases reproduced from Fig. 7 in [Morigaki *et al.*, 2005a], that is, thermal annealing in air and in vacuum of light-induced dangling bonds in pm-Si:H by Morigaki *et al.* [2005a] and thermal annealing of light-induced dangling bonds in a-Si:H by Jackson and Kakalios [1988], respectively [Takeda *et al.*, 2008].

Thermal annealing of dangling bonds created during deposition, that is, native dangling bonds in sputtered a-Si and a-Si:H, is interesting in the temperature dependence of β. As shown in Fig. 3.95, the values of β for sputtered a-Si:H prepared at room temperature with various hydrogen content [H] (12 at.%–38 at.%) are scattered in the range of 0.4–1.0. For sputtered a-Si, the temperature dependence of β is opposite to that for a-Si:H. This seems to be related to the annihilation of voids, which proceeds when the annealing temperature is increased. According to Schlesinger and Montroll [1984], the relaxation process in a fractal-time random walk is expressed as a stretched exponential function in which β corresponds to the fractal dimension. The fractal dimension associated with voids seems to decrease when the voids are annealed out as the temperature increases that is consistent with the observed temperature dependence of β. This is just a qualitative speculation for the dispersive behavior of the annealing of dangling bonds; a quantitative explanation is required to understand fully the annealing behavior in a-Si.

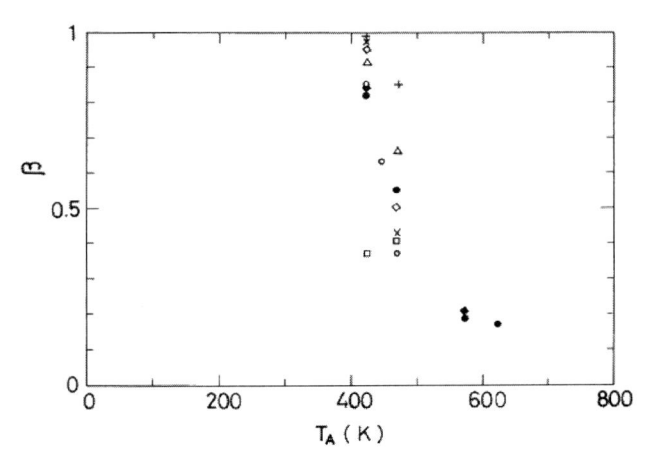

Figure 3.95 Plots of β *vs.* T_A for a-Si and a-Si:H. For a-Si, film thickness (nm): ◆: 150 and ●: 2300. For a-Si:H, hydrogen content (at.%): + : 12, × : 16, ○: 20, ◇: 26, △: 30, and □: 38. [Reproduced from Fujita *et al.*, *Philos. Mag. B*, **69**, 57 (1994) by permission of Taylor & Francis.]

Chapter 4

Hydrogenated Microcrystalline Silicon

4.1 Introduction

As mentioned in Section 3.1, hydrogenated microcrystalline silicon films (μc-Si:H) are prepared, using PECVD with a mixture gas of SiH_4 and H_2, whose gas dilution ratio, γ, is beyond a certain critical value γ_c (= 3.5% for T_S = 100°C). These films are also prepared, using HWCVD, in which γ_c is 8% for T_S = 230°C. The film consists of two phases, that is, microcrystalline and amorphous phases. The crystalline fraction measured from Raman scattering, as mentioned in Section 3.1, increases with decreasing γ or with increasing hydrogen dilution, D_H ($\equiv 1 - \gamma$), as shown in Fig. 4.1 for HWCVD films [Niikura *et al.*, 2011].

The properties of defects in μc-Si:H have been investigated by measurements of ESR, photoconductivity (PC), PL, ODMR, and so on. In Section 4.2, ESR properties of defects are presented. Light-induced defects, and PL and ODMR measurements are described in Sections 4.3 and 4.4, respectively. In Section 4.5, light-induced effects on PL are treated.

Light-Induced Defects in Semiconductors
Kazuo Morigaki, Harumi Hikita, and Chisato Ogihara
Copyright © 2015 Pan Stanford Publishing Pte. Ltd.
ISBN 978-981-4411-48-6 (Hardcover), 978-981-4411-49-3 (eBook)
www.panstanford.com

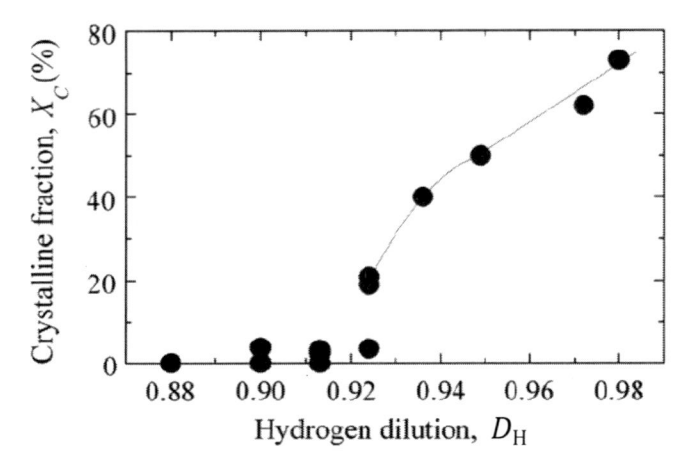

Figure 4.1 Crystalline fraction X_c deduced from Raman scattering analysis as a function of hydrogen dilution D_H. [Reproduced from Niikura *et al*, *Thin Solid Films*, **519**, 4502 (2011) by permission of Elsevier.]

4.2 Defects

The principal defect in µc-Si:H is an Si-dangling bond. The ESR spectrum of dangling bonds (DBs) in HWCVD samples has an asymmetric lineshape, as* shown in Fig. 4.2. This shape can be fitted by assuming anisotropic g-values, as seen in the figure, in which a Gaussian spin-packet is taken, and the following values of g-values, $g_{//}$ and $g_\perp$, and standard deviations, $\sigma_{//}$ and $\sigma_\perp$, are obtained from the best fitting [Morigaki *et al.*, 2002]: $g_{//}$ = 2.0014, $g_\perp$ = 2.0068, $\sigma_{//}$ = 3.2 G, and $\sigma_\perp$ = 4.4 G. Such a model of anisotropic magnetic centers for DBs in µc-Si:H has already been suggested by Kondo *et al.* [2000] and Ehara *et al.* [2000].

Kondo *et al.* [2000] made an analysis of observed ESR spectra, assuming that there exist anisotropic DB centers with an axially symmetric axis along [111] in [111]-oriented crystallites, randomly oriented DB centers (powder average), and DB centers in the amorphous phase with weight of 0.6, 0.05, and 0.35, respectively, because they observed different spectra when magnetic field is applied parallel and perpendicular to the substrate of samples. Their obtained g-values ($g_{//}$ = 2.0022, $g_\perp$ = 2.0078) are close to those of P_b centers ($g_{//}$ = 2.0023, $g_\perp$ = 2.0078) observed at an interface between c-Si and its surface oxide layer [Nishi, 1971], so that they have suggested that the DB center is similar to the P_b center (both centers have an axially symmetric axis along [111]).

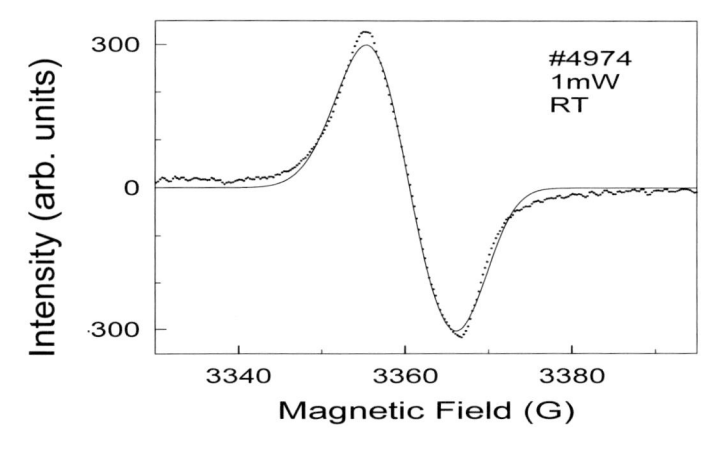

Figure 4.2   The observed ESR spectrum of μc-Si:H No. 4974 ($X_c = 0.81$) at room temperature. The fitted curve in terms of the anisotropic magnetic center model is shown by a solid curve. See the text for the values of the fitting parameters. [Reproduced from Morigaki *et al.*, *J. Non-Cryst. Solids*, **299–302**, 561 (2002) by permission of Elsevier.]

In our case [Morigaki *et al.*, 2002], although slightly different ESR spectra have been observed when the axis of the substrate of samples is rotated about the magnetic field, differences are not large compared with those of Kondo *et al.* [2000].

On the other hand, Finger *et al.* [2005] and Neto *et al.* [2004] interpreted their observed ESR spectra in terms of two components, that is, the existence of two different types of DBs in μc-Si:H, that is, DBs in the column boundary and oxygen-related DBs in the column boundary. Their samples were prepared by PECVD. Lips *et al.* [2003] have pointed out that this model is consistent with different microwave power saturation behaviors of two components and their different changes associated with the doping level. However, the spectrum can be interpreted in terms of an asymmetric magnetic center model, as shown in Fig. 4.3. The values of the fitting parameters are given as follows: $g_{//} = 2.00375$, $g_{\perp} = 2.00545$, $\sigma_{//} = 1.8$ G, and $\sigma_{\perp} = 5.2$ G.

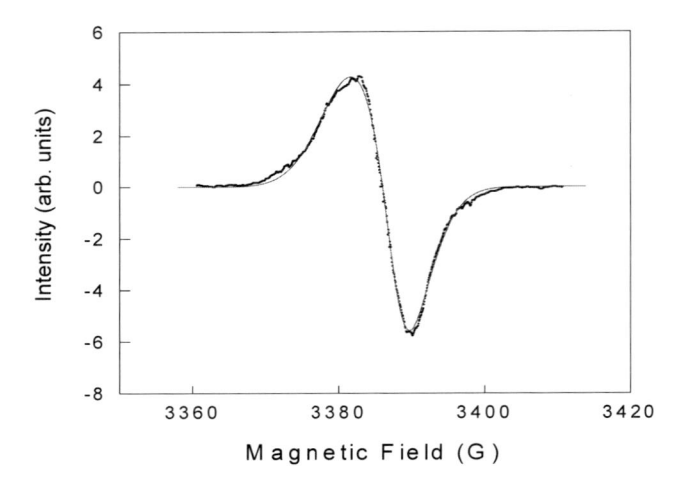

Figure 4.3 The observed ESR spectrum (the dotted curve) of μc-Si:H (type I) (X_c = 0.82) by Finger *et al.* [2005]. The fitted curve in terms of the anisotropic magnetic center model is shown by a solid curve. [Reproduced from Morigaki *et al.*, *Kotai Butsuri (Solid State Physics)*, **44**, 211 (2009c) by permission of AGNE Gijutsu Center.]

The g-values, $g_{//}$, and $g_\perp$, are plotted as functions of crystalline fraction X_c in HWCVD samples and PECVD samples, as shown in Figs. 4.4 and 4.5, respectively. The g-values observed in μc-Si:H have already been discussed in Section 3.2.2. These results are discussed on the basis of calculation by Ishii *et al.* [1981] and Ishii and Shimizu [1998]. They calculated the values of three principal components of the g-tensor, g_1, g_2, and g_3, as functions of angle θ between the DB and its back-bond, using extended Hückel theory and iterative extended Hückel theory, as shown in Fig. 4.6. In our model, g_3 corresponds to $g_{//}$, and g_1 and g_2 correspond to $g_\perp$ ($\equiv g_1 = g_2$). First, we consider the case of HWCVD samples in Fig. 4.4. The observed value of $g_\perp$ is close to $g_\perp$ at = 102.5°, that is, smaller than the tetrahedral angle. From the experimental result shown in Fig. 4.4, variation of $g_\perp$ with X_c is small and the observed value of $g_\perp$ is smaller than the calculated value at the tetrahedral angle in all the measured range. Such a variation of $g_\perp$ with X_c seems to be related to the size of crystallites. This has been discussed in terms of a surface tension model, assuming that DB is located on the surface of crystallites [Morigaki and Niikura, 2005].

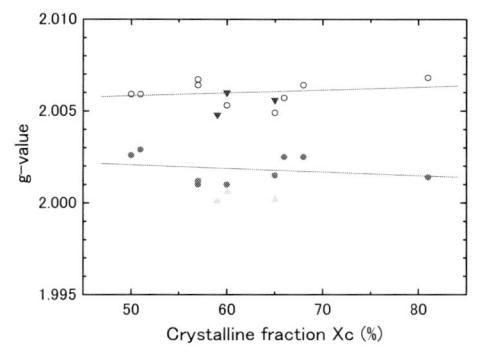

Figure 4.4 g-values, $g_{//}$ and $g_\perp$, measured in HWCVD µc-Si:H *vs.* crystalline fraction, χ_c. Experimental points [Morigaki *et al.*, 2002] are classified by two groups of HWCVD µc-Si:H samples prepared at 230–250°C and 200°C. For the former samples, closed circles and open circles represent $g_{//}$ and $g_\perp$, respectively. For the latter samples, closed triangles and closed inverse-triangles represent $g_{//}$ and $g_\perp$, respectively. The two lines are the least-square fits for experimental points of $g_{//}$ and $g_\perp$ for the former samples [Morigaki and Niikura, 2005].

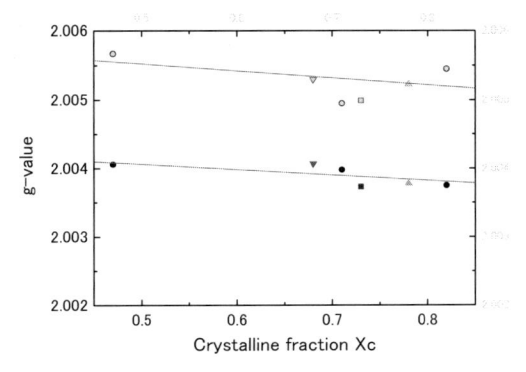

Figure 4.5 g-values, $g_{//}$ and $g_\perp$, estimated in terms of the anisotropic magnetic center model for the observed ESR spectra of µc-Si:H by Finger *et al.* [2005] and Neto *et al.* [2004] *vs.* crystalline fraction, X_c. For the VHF PECVD films [Finger *et al.*, 2005], closed circles and open circles represent $g_{//}$ and $g_\perp$, respectively. For the RF PECVD films [Neto *et al.*, 2004], closed squares and open squares represent $g_{//}$ and $g_\perp$, respectively. For the HWCVD films [Neto *et al.*, 2004], closed inverse triangles and open inverse triangles represent $g_{//}$ and $g_\perp$, respectively. The two lines are the least-square fits for experimental points of $g_{//}$ and $g_\perp$ for these samples [Morigaki *et al.*, 2009c].

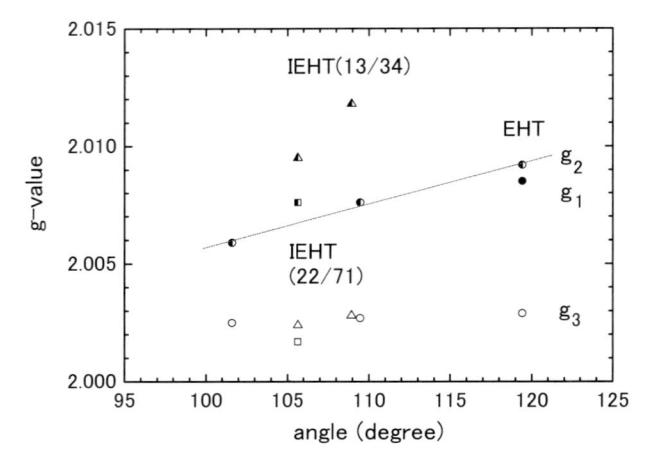

Figure 4.6 Calculated orthorhombic components of g-tensor of dangling bonds, g_1, g_2, and g_3, *vs.* angle θ, between the orientations of the dangling bond and its Si-Si backbond [Isshi *et al.*, 1981; Ishii *et al.*, 1998]. EHT: Extended Hückel theory, IEHT: Iterative extended Hückel theory. The symbol (k/n) means that the calculation was carried out for the n-atom Si cluster and then only the orbitals centered on k Si atoms including the defect atom(s) of the wave functions were taken into account in g-values [Ishii *et al.*, 1998]. The symbols of circles, squares, and triangles are calculated g-values on the basis of EHT, IEHT (22/71), and IEHT (13/34), respectively. g_1, g_2, and g_3 are shown by closed symbols, semi-closed symbols, and open symbols, respectively. A line represents the least-square fit for calculated points of g_2 by EHT [Morigaki and Niikura, 2005].

For PECVD samples, on the contrary, the variation of $g_\perp$ with X_c is opposite to that for HWCVD samples, as seen in Fig. 4.5. Such a difference in the variation of $g_\perp$ with X_c for two types of samples seems to be related with a difference in their morphology of amorphous and microcrystalline phases from the viewpoint of the surface tension model [Morigaki and Niikura, 2005].

The PECVD samples prepared at 100°C show different features in g-values and spin densities between $X_c < 0.3$ and $X_c > 0.5$, as shown in Figs. 4.7 and 4.8. The g-values, $g_{//}$ and $g_\perp$, for $X_c > 0.5$ are close to those of the positively charged vacancy (positive correlation energy) in c-Si, so that the magnetic center may be identified as being such a defect inside the crystallite (see Fig. 3.9). This is consistent with the role of nonradiative centers for this center. For $X_c < 0.3$, the magnetic center is a DB on the surface of crystallites, as mentioned before.

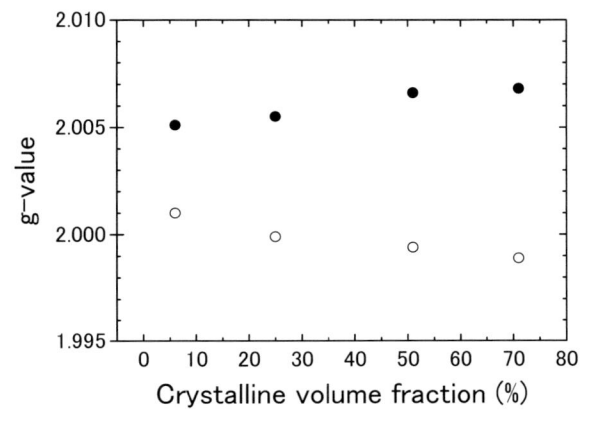

Figure 4.7 g-values, $g_{//}$ and $g_\perp$, measured in µc-Si:H *vs.* crystalline volume fraction, X_C. The open and closed circles designate $g_{//}$ and $g_\perp$, respectively. [Reproduced from Morigaki *et al.*, *J. Appl. Phys.*, **105**, 083703-1 (2009b) by permission of American Institute of Physics.]

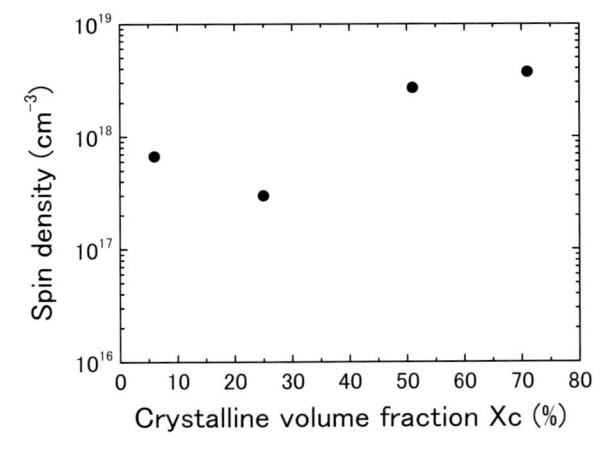

Figure 4.8   Spin densities *vs.* crystalline volume fraction, X_c. [Reproduced from Morigaki *et al.*, *J. Appl. Phys.*, **105**, 083703-1 (2009b) by permission of American Institute of Physics.]

For those samples, the conduction electron ESR has been observed at room temperature, as shown in Fig. 4.9. This ESR line has also been observed at room temperature for some of PECVD samples prepared at 200–250°C [Finger *et al.*, 1994].

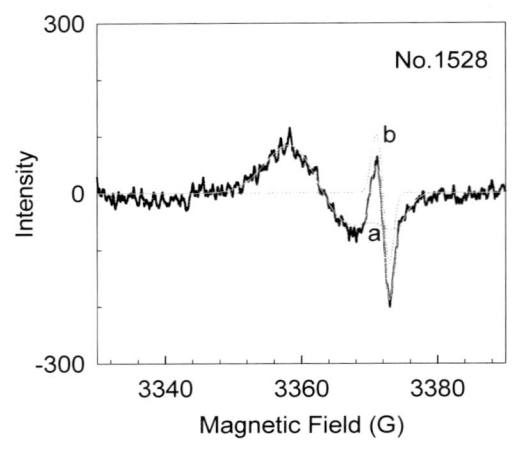

Figure 4.9 The observed ESR spectrum of samples nos. 1528 and fit of the observed ESR spectrum by two lines due to anisotropic magnetic centers and conduction electrons for sample no. 1528. The solid curve is superposition of calculated curves a and b, that is, dashed curves a (defects) and b (conduction electrons). [Reproduced from Morigaki *et al.*, *J. Appl. Phys.*, **105**, 083703-1 (2009b) by permission of American Institute of Physics.]

4.3 Light-Induced Defects

Under illumination, a new ESR line at g = 1.998 with N_S = 3 × 10^{16} cm^{-3} has been observed at 10 K in PECVD μc-Si:H samples by Kondo *et al.* [1998], as shown in Fig. 4.10. This LESR signal intensity was measured as a function of photon energy, as shown in Fig. 4.11. As seen in this figure, the intensity increases above 1.1 eV corresponding to the band gap energy of c-Si with increasing photon energy. Its temperature dependence was also measured, as shown in Fig. 4.12. Above 40 K, it decreases exponentially with increasing temperature (activation energy: 11 meV) and disappears above 100 K, as shown in Fig. 4.12. At lower temperature the spin density follows the Curie law, that is, $\propto T^{-1}$. From these results, Kondo *et al.* [1998] have concluded that the magnetic center responsible for the g = 1.998 signal arises from a localized electron trapped at a shallow level (tail electron) close to the edge of the conduction band with a depth of 11 meV in the crystallites.

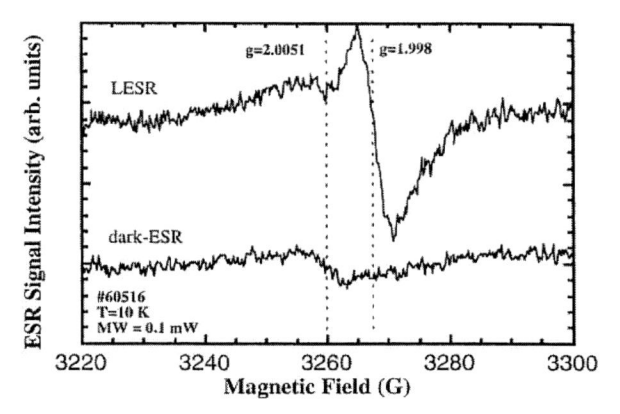

Figure 4.10   LESR (upper) and dark-ESR (lower) spectra for μc-Si:H prepared at 300°C. In the dark, only the dangling bond signal (g = 2.0051) is observed. Under illumination, additional component with g = 1.998 appears. The spin density of the LESR center (C-center) is estimated to be $3 \times 10^{16}\,cm^{-3}$. [Reproduced from Kondo *et al.*, *J. Non-Cryst. Solids*, **227–230**, 1031 (1998) by permission of Elsevier.]

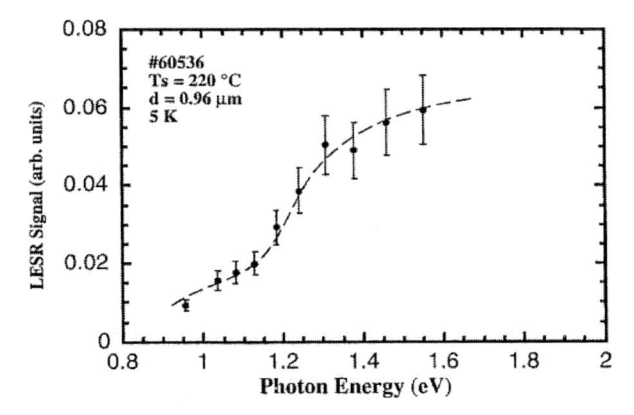

Figure 4.11   Excitation photon energy dependence of the LESR response in μc-Si:H. The measurement was done at 5 K using halogen lamp with a monochrometer. The LESR response is corrected by the spectral dependence of the excitation intensity. The dashed line is a guide for the eye. [Reproduced from Kondo *et al.*, *J. Non-Cryst. Solids*, **227–230**, 1031 (1998) by permission of Elsevier.]

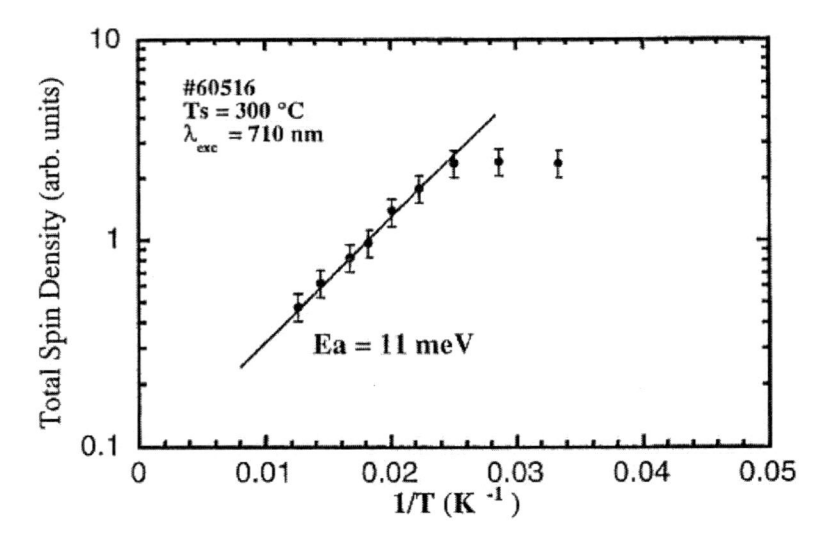

Figure 4.12 Measuring temperature dependence of the total spin density, which is the ESR spin density multiplied by T. The excitation wavelength is 710 nm. The density starts to decrease at 40 K and the signal disappears for $T > 100$ K. The activation energy is estimated to be 11 meV. The solid line designates a fitted result. [Reproduced from Kondo *et al.*, *J. Non-Cryst. Solids*, **227–230**, 1031 (1998) by permission of Elsevier.]

As seen in Fig. 4.10, a DB ESR line is enhanced under illumination compared with before illumination (in the dark). Kondo *et al.* [1998] conclude that such a light-induced DB formation is related to surface oxidation of the sample.

de Lima *et al.* [2002] observed an ESR signal at low temperatures under illumination, as shown in Fig. 4.13. The signal is best described by two powder patterns, that is, an asymmetric line at $g_{//} = 1.999$, and $g_{\perp} = 1.996$ and a broad line at $g = 1.998$. They have suggested that these centers are a conduction band tail electron and a valence band tail hole, respectively. Su *et al.* [2008] observed a narrow line at $g = 1.996$ and a broad line at $g = 1.998$ in PECVD samples with average size of crystallites of 20 nm ($X_c \cong 0.50$) under illumination, as shown in Fig. 4.14. The relative intensities of two signals depend on the excitation source, that is, 1:2 for 810 nm LED, 1:1.3 for pulsed YAG laser, and 1 : 1 for CW-YAG laser.

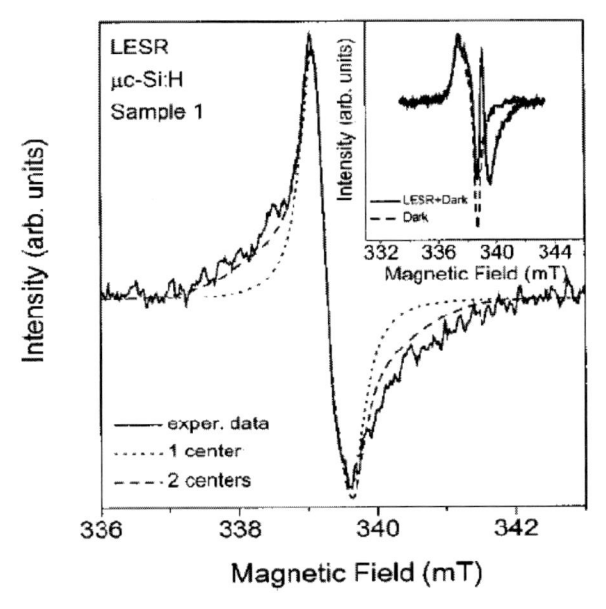

Figure 4.13 LESR signal of μc-Si:H taken at 15 K using a microwave power of 10 mW and a modulation amplitude of 1 G at a microwave frequency of 9.489 GHz. The dotted line is the result of a simulation using a powder pattern with $g_{//}$ = 1.999 and $g_\perp$ = 1.996 and Lorentzian broadening of 3 G. The dashed line represents a simulation in which a second center with g = 1.998 is also included with an equal concentration. The inset shows both the dark signal and the (LESR + dark) signal. [Reproduced from de Lima *et al.*, *Phys. Rev. B*, **65**, 235324-1 (2002) by permission of American Physical Society.]

The LESR signal observed in μc-Si:H (undoped, n-type, and p-type) under illumination by Finger *et al.* [1994, 1998] is shown in Fig. 4.15, in which the observed LESR spectrum consisting of two components at g = 2.0043 and g = 2.01 in a-Si:H is also shown as well as dark ESR signals in μc-Si:H and a-Si:H. The two components in μc-Si:H being attributed to photoexcited conduction electrons and DB electrons arising from excitation from D^- states or trapping of holes in D^- states are observed at g = 1.9983 and g = 2.0052, respectively, under illumination. Finger *et al.* [1998] measured the LESR signal decay curve and further responses for pulsed IR illumination. The spin-lattice relaxation time T_1 is also measured as a function of temperature, using a pulsed technique, as shown in Fig. 4.16.

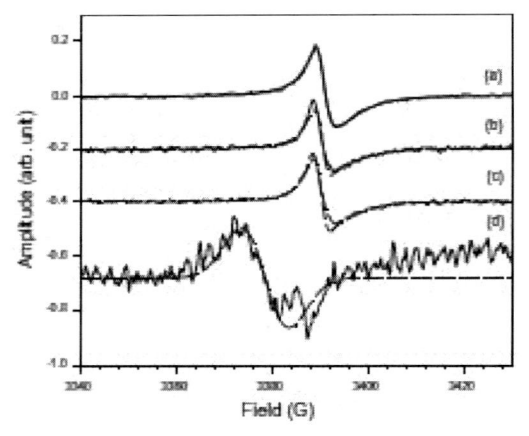

Figure 4.14   LESR signal in nc-Si:H at 7 K with illumination with an 810 nm LED (a), a pulsed YAG laser (b), a CW YAG laser (c), and a-Si: H illuminated with a CW YAG laser at 7 K (d). The solid lines represent data: the dashed lines represent the fits to the data. The small feature near 3385 G in (d) is due to E' centers at the film/substrate interface. [Reproduced from Su *et al., J. Non-Cryst. Solids*, **354**, 2231 (2008) by permission of Elsevier.]

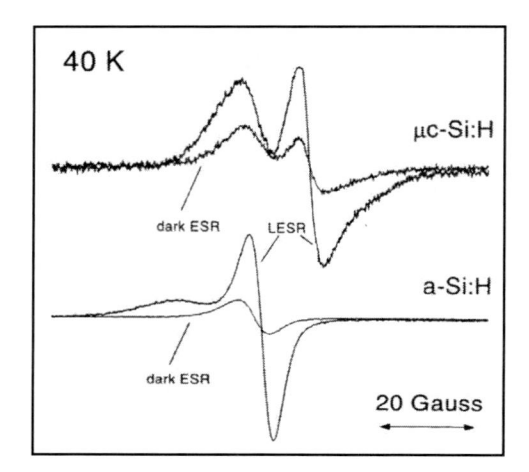

Figure 4.15   LESR together with dark ESR spectra of an n-type µc-Si:H sample, in comparison with a high-quality (spin density = 4 × 10^{15} cm^{-3}) a-Si:H sample. [Reproduced from Finger *et al., Philos. Mag. B*, **77**, 805 (1998) by permission of Taylor & Francis.]

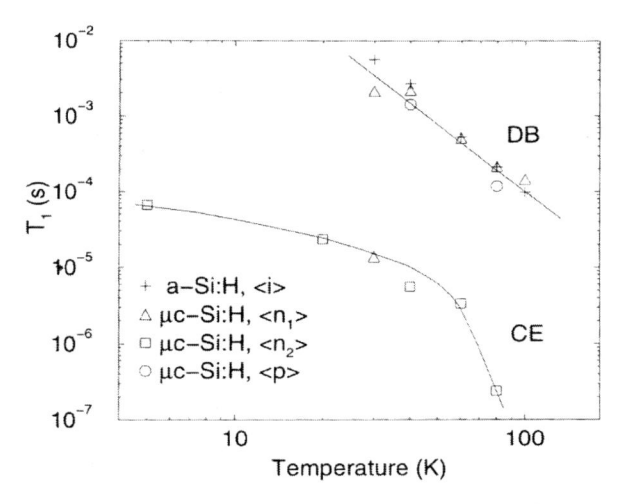

Figure 4.16 Spin-lattice relaxation times T_1 of the conduction and the dangling bond resonances in μc-Si:H as functions of temperature for samples with different doping levels. For comparison, the T_1 relaxation times of the dangling bond resonance in a-Si:H are also shown. [Reproduced from Finger *et al.*, *Philos. Mag. B*, **77**, 805 (1998) by permission of Taylor & Francis.]

4.4 Photoluminescence and Optically Detected Magnetic Resonance

The PL spectra observed at 7 K in PECVD μc-Si:H samples with various values of X_c prepared at 100°C are shown in Fig. 4.17 [Yamaguchi and Morigaki, 1993]. A low-energy part of PL arises from the crystalline phase, while the PL above 1.0 eV arises from the amorphous phase. The ODMR measurements have been performed by Malten *et al.* [1998], Depinna *et al.* [1983], and Boulitrop *et al.* [1983]. Malten *et al.* [1998] observed an enhancing signal at 10 K at g = 1.998 (FWHM = 16–18 G) and a quenching signal at g = 2.005 (FWHM = 30 G), using microwave-modulation frequency of 1.4 kHz, and a broad enhancing signal deconvoluted into two components at g = 2.08 (FWHM = 200 G) and at g = 2.13 (FWHM = 1000 G), and a quenching signal at g = 2.005, using microwave-modulation frequency of 2.4 kHz, as shown in Figs. 4.18 (a) and (b), respectively. They interpreted that an enhancing signal at g = 1.998 is due to conduction band tail states or impurity-related donor states and that a quenching signal is due to DBs existing in grain boundaries. The origin of the lines at g = 2.13 and g = 2.08 is recombination involving states near or in the valence band.

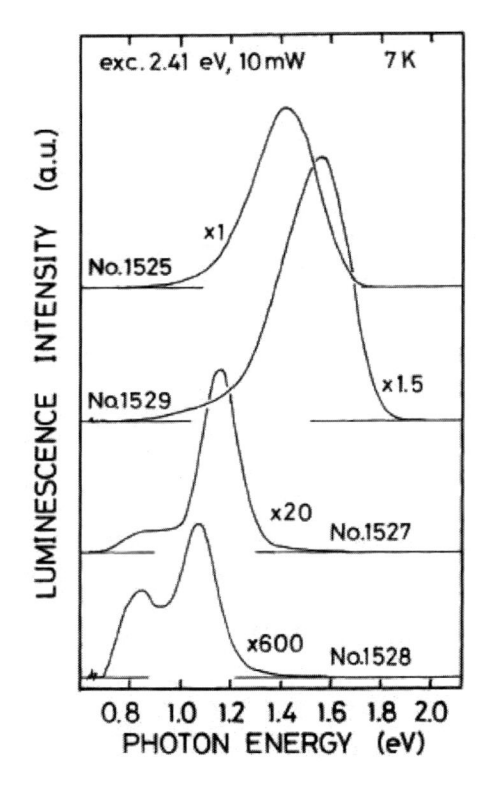

Figure 4.17 Steady-state PL spectra taken at 7 K for a-Si:H (sample no. 1525) and μc-Si:H (samples nos. 1527 (X_c = 0.51), 1528 (X_c = 0.71), and 1529 (X_c = 0.06)). [Reproduced from Yamaguchi and Morigaki, *J. Phys. Soc. Jpn.*, **62**, 2915 (1993) by permission of The Physical Society of Japan.]

Depinna *et al.* [1983] observed two enhancing components at g = 1.989 (FWHM = 220 G) and g = 2.006 (FWHM = 590 G), monitoring PL below 1.2 eV and a quenching component at g = 2.0050, monitoring PL above 1.0 eV, as shown in Fig. 4.19. Boulitrop *et al.* [1983] observed an enhancing line at g = 1.9997 (FWHM = 18.5 G), monitoring PL below and at 0.75 eV and a quenching line at

g = 2.0043 (FWHM = 25 G), monitoring PL at 0.85 eV, as shown in Fig. 4.20.

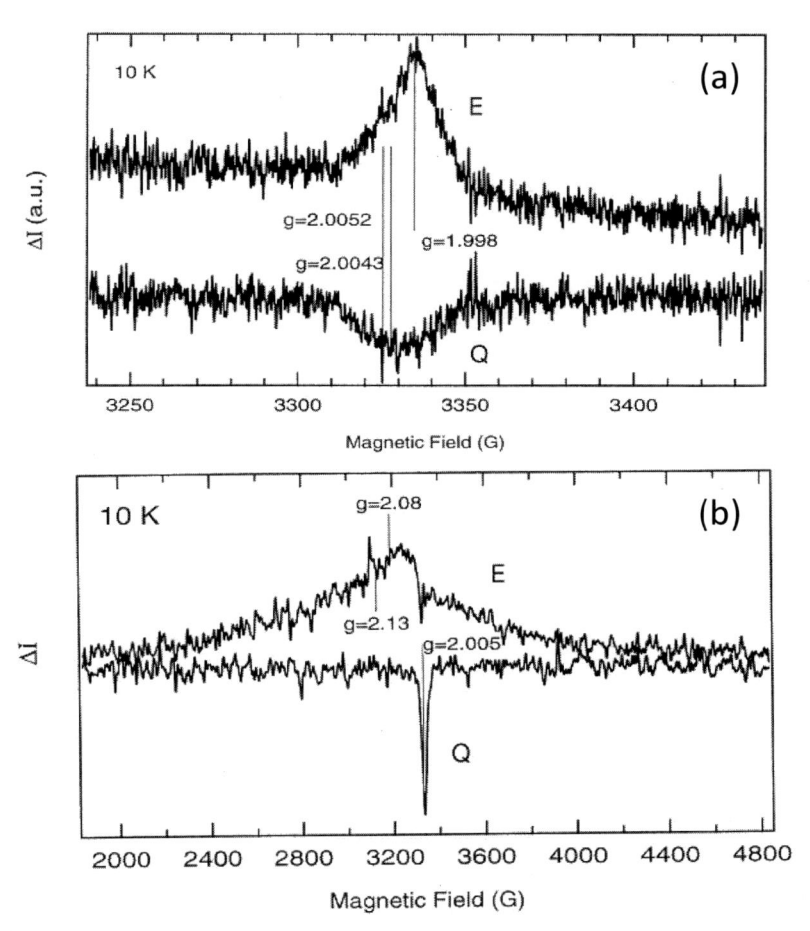

Figure 4.18 (a): ODMR spectra of an undoped µc-Si:H film, sweep width 200 G, microwave modulation frequency 1.4 kHz. Markers show the field values corresponding to g-values of g = 2.0043, g = 2.0052 (dangling bonds), and g = 1.998 (conduction electrons). E: enhacing, Q: quenching resonance. (b) ODMR spectra of undoped µc-Si:H films measured at 10 K and a modulation frequency of 2.4 kHz. Broad and narrow lines can be distinguished clearly. [Reproduced from Malten *et al.*, *MRS Proc.*, **507**, 787 (1998) by permission of Materials Research Society.]

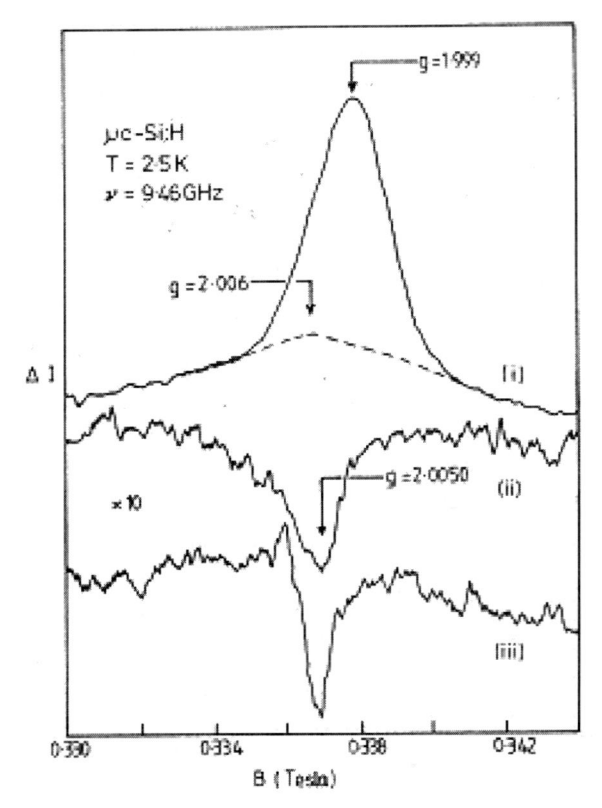

Figure 4.19 ODMR spectra of μc-Si:H with the Ge detector by monitoring the emission (i) $E < 1.2$ eV, (ii) and (iii) $E > 1.0$ eV. Different phase conditions account for the difference between the curves (ii) and (iii) which are typical for a-Si:H. [Reproduced from Depinna *et al.*, *Philos. Mag. B*, **47**, L57 (1983) by permission of Taylor & Francis.]

Concerning the identification of an enhancing line at $g = 1.998$ and of a quenching line at $g = 2.005$, these three groups have a similar conclusion, but concerning other lines (due to hole states), further investigations are required. From the ODMR measurements, it is concluded that conduction band tail electrons participate in radiative recombination responsible for principal PL in μc-Si:H and that the DBs in the grain boundary are nonradiative recombination centers.

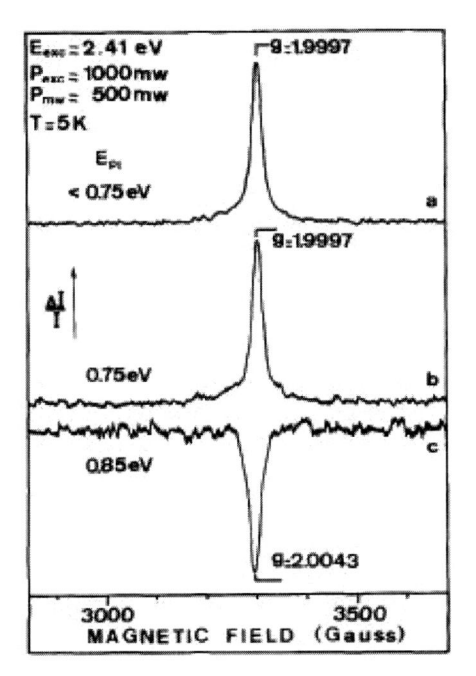

Figure 4.20 ODMR spectra of a post-hydrogenated sample of μc-Si:H, recorded at PL energies (a) less than 0.75 eV, (b) at 0.75 eV, and (c) at 0.85 eV. [Reproduced from Boulitrop *et al.*, *Solid State Commun.*, **48**, 181 (1983) by permission of Elsevier.]

4.5 Light-Induced Effects on Photoluminescence

The light-induced effect on PL in μc-Si:H is shown in Fig. 4.21, in which the PL spectra of a-Si:H and μc-Si:H are taken before and after illumination by an argon-ion laser of 514.5 nm (2.41 eV) at light intensity of 1.2 W/cm^2 for 120 min. The light-induced effect on PL in a-Si:H is mentioned in Section 3.4.3. On the contrary, in μc-Si:H, the PL intensity decreases after illumination compared with before illumination, but it is very small. Such changes in the PL intensity before and after illumination, I_B and I_A, at the peak of the spectrum, E_p, are shown in Fig. 4.22. From these results, it is concluded that μc-Si:H is rather stable against light-induced degradation compared with a-Si:H.

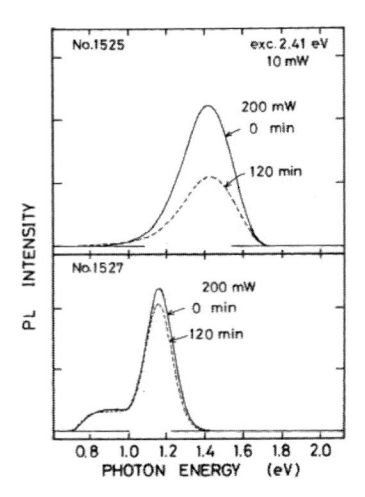

Figure 4.21 The illumination-time dependence of PL spectra taken after strong illumination (intensity = 200 mW) as well as before illumination for a-Si:H (sample No.1525) and μc-Si:H (sample no. 1527). [Reproduced from Yamaguchi and Morigaki, *J. Phys. Soc. Jpn.*, **62**, 2915 (1993) by permission of The Physical Society of Japan.]

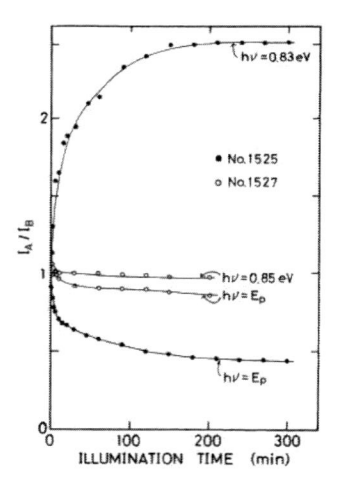

Figure 4.22 I_A/I_B vs. illumination (intensity = 200 mW) time at an excitation intensity = 10 mW for PL measurements for a-Si:H (sample no. 1525, solid circles) and μc-Si:H (sample no. 1527, open circles). [Reproduced from Yamaguchi and Morigaki, *J. Phys. Soc. Japan*, **62,** 2915 (1993) by permission of The Physical Society of Japan.]

Chapter 5

Amorphous Chalcogenides

5.1 Introduction

Recently, amorphous chalcogenides (a-Chs) have been paid much attention from their applications, particularly X-ray image detectors and optical memories (phase-change memories), for example, digital versatile disks (DVDs). The change in the refractive index of this material from an amorphous to crystalline state is used for optical memory disks. From the viewpoint of physics, this material is interesting because the electron-phonon interaction is strong. In this chapter, we deal with typical light-induced phenomena and the nature of light-induced defects in a-Chs. For further reading including on electronic properties of a-Chs and their preparation, the following book is recommended: Keiji Tanaka and Koichi Shimakawa [2011].

5.2 Amorphous Chalcogenides

Amorphous chalcogenides (a-Chs) exhibit light-induced phenomena, particularly for reversible changes associated with illumination,

Light-Induced Defects in Semiconductors
Kazuo Morigaki, Harumi Hikita, and Chisato Ogihara
Copyright © 2015 Pan Stanford Publishing Pte. Ltd.
ISBN 978-981-4411-48-6 (Hardcover), 978-981-4411-49-3 (eBook)
www.panstanford.com

the following three phenomena are remarkable: photodarkening, volume changes, and defect creation. In the following, these are briefly summarized.

(1) Photodarkening (PD)

The optical absorption edge shifts toward lower energy (red-shift) under higher energy illumination (e.g., 2.41 and 2.54 eV), which means a decrease in the optical bandgap E_0, $\Delta E_0/E_0 \cong -2$ % for amorphous As_2S_3 (a-As_2S_3), and subgap illumination of 1.92 eV recovers particularly the red-shift, as shown in Fig. 5.1. The PD effect depends on temperature, that is, ΔE_0 decreases with decreasing temperature toward the glass transition temperature T_g. The PD effect is annealed out at T_g, as shown in Fig. 5.2.

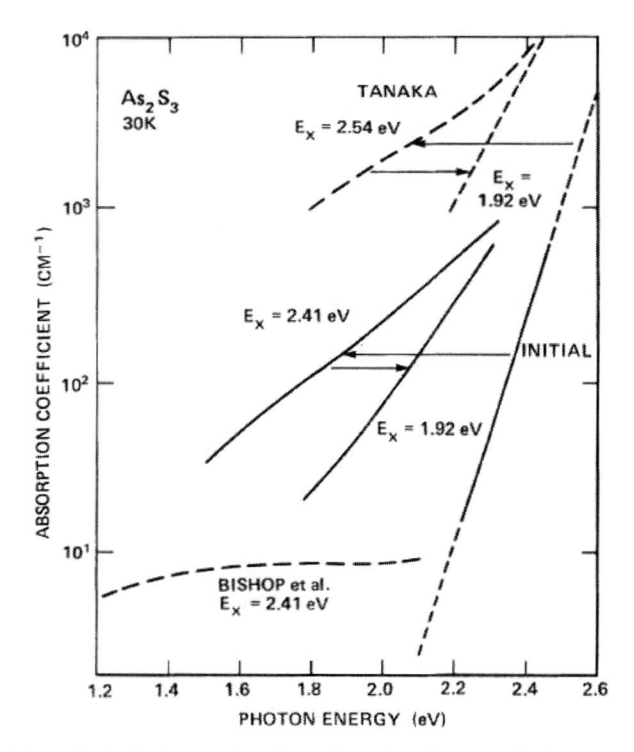

Figure 5.1   Optical absorption data showing the initial absorption edge, the result of light exposure, and of subsequent bleaching. [Reproduced from Biegelsen and Street, *Phys. Rev. Lett.*, **44**, 803 (1980) by permission of American Physical Society.]

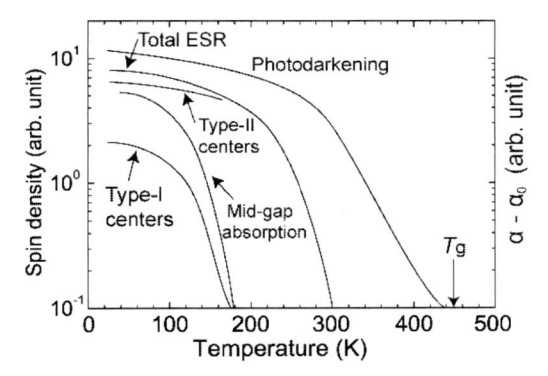

Figure 5.2   Annealing behaviors of photoinduced ESR (Total, type I, and type II), mid-gap absorption, and photodarkening in glassy As_2Se_3. The data are based on the results by Hautala *et al.* (1988) and Tanaka and Shimakawa (2011).

(2) Photo-induced volume changes

The photo-induced volume changes (PVCs) are normally volume expansion, for example, $\Delta V/V \cong 6\%$ for obliquely deposited a-As_2S_3 films, and always occur with PD. An example of PVC is shown in Fig. 5.3, in which relative changes in thickness $\Delta d/d$ in obliquely deposited a-As_2S_3 films are shown with illumination time and time after illumination is switched off. The PVC effect can be recovered by annealing at T_g.

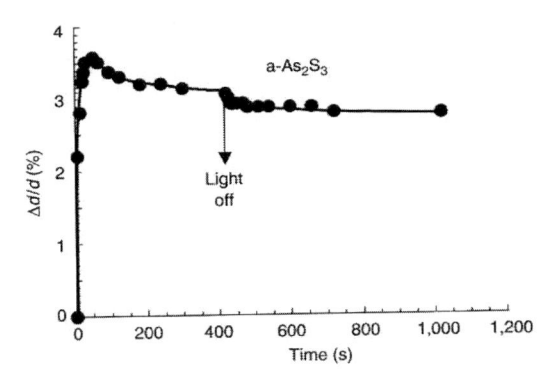

Figure 5.3   Relative changes in thickness, $\Delta d/d$, with time for obliquely deposited (80°C) amorphous As_2Se_3 (a-As_2Se_3) films. Line is drawn as a guide for the eye. [Reproduced from Ganjoo *et al.*, *J. Non-Cryst. Solids*, **266–269**, 919 (2000) by permission of Elsevier.]

The theoretical investigations on the PVC have been performed by Lukács *et al.* [2010] and Lukács and Kugler [2011].

(3) Photo-induced defect creation (PDC)

The microscopic nature of defects is generally elucidated from magnetic resonance techniques such as ESR, ENDOR, ODMR, ODENDOR, and so on, similar to those in a-Si:H. No ESR signals are normally observed in the dark in a-Chs., but ESR signals under illumination, that is, light-induced ESR (LESR) signals have been observed, as shown in Fig. 5.4 for three kinds of a-Chs, that is, a-As$_2$S$_3$, a-As$_2$Se$_3$, and amorphous selenium (a-Se). The ESR spectra consist of two types of LESR centers classified by thermal stability, that is, the type I center is less thermally stable (completely annealed at ~190 K), whereas the type II center is more thermally stable (annealed at room temperature), as shown by their thermal annealing measurements in Fig. 5.2. Each center has two components due to electrons and holes. The type I and II centers constitute approximately 15% and 85% of the induced spins, respectively, after prolonged illumination at high intensities (≥100 mW/cm^2) at low temperatures. For the origins of two types of LESR centers

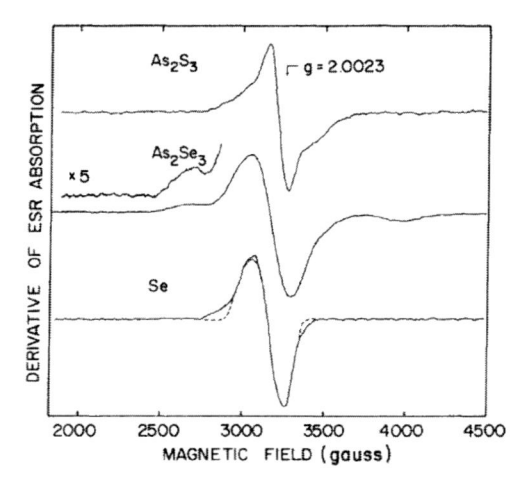

Figure 5.4   Optically induced ESR spectra obtained near 4.2 K in chalcogenide glasses. Dashed line superimposed in Se curve is a computer simulation. [Reproduced from Bishop *et al.*, *Phys. Rev. B*, **15**, 2278 (1977) by permission of American Physical Society.]

in a-As_2S_3, Hautala *et al.* [1988] suggested that the type I centers are an electron predominantly localized on a s-p hybridized orbital of a twofold-coordinated arsenic atom (As_I) (more delocalized than As_{II}) and a hole on a nonbonding 3p orbital of a probably singly coordinated sulfur atom (S_I), while the type II centers are an electron on a nonbonding 4p orbital of a twofold-coordinated arsenic atom (As_{II}) and a hole predominantly localized on a nonbonding 3p orbital of a sulfur atom (S_{II}). These As_{II} and S_{II} are created by the breaking of As–As bonds and S–S bonds, respectively. This is based on their result that the relative spin density of the As_{II} center decreases rapidly with increasing arsenic content x, whereas that of the S_{II} center increases with increasing x. These models of centers stand on a model of D^+–D^- pairs proposed by Biegelsen and Street, in which D^+ and D^- designate positively charged defects and negatively charged defects, respectively. D^+ and D^- correspond to C_3^+ and C_1^+ in a-Se, respectively, as shown in the atomic configuration in Fig. 5.5, in which the superscript and

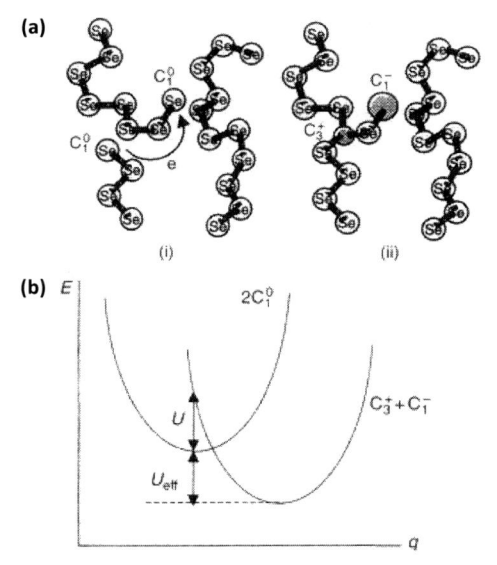

Figure 5.5 (a) Formation of charged defects (valence alternation pairs: VAP) in a-Se. (b) Configuration coordinate diagram for the formation of a $D^+(C_3^+)$ – $D^-(C_1^-)$ pair. The overall energy is lowered by the effective correlation energy U_{eff}. [Reproduced from Singh and Shimakawa, *Advances in Amorphous Semiconductors* (2003) by permission of Taylor & Francis.]

the subscript designate the charge state of defects and their coordination number, respectively. These defects are stabilized by the effective correlation energy U_{eff}, as shown in Fig. 5.5(b). Such a stabilization is caused by strong electron-phonon interaction [Street and Mott, 1975]. The D^+-D^- pair is called the valence-alternation pair [Kastner *et al.*, 1976]. At high-intensity illumination above 100 mW/cm^2, the inducing spin density does not saturate with increasing illumination time and exceeds 10^{20} cm^{-3}. At pre-existing defects, D^+ and D^- limit the inducing spin density, $\sim 10^{17}$ cm^{-3}; this means that more defects are created by prolonged illumination. These defects are mostly annealed out at T_g.

Elliott and Shimakawa proposed a mechanism for breaking of bonds to create light-induced defects: for example, for a-As$_2$S$_3$, As–S bonds are broken and, as a consequence, the type I LESR centers are an electron at As$_1^+$(2S) and a hole at C$_1^-$(As), in which the symbol in the parentheses denotes the neighboring atom to the defect. The type II LESR centers are an electron at As$_2^+$(As, S) and a hole at C$_1^-$(S), as illustrated in Figs. 5.6(a)–(c). As shown in Figs. 5.6(b) and (c) for n-type II center, there is a wrong bond, either an As–As bond or a S–S bond neighboring to the As–S bond to be broken by illumination. From the result, the model is qualitatively consistent with the result of the relative spin density of the type I and type II centers as a function of arsenic content measured by Hautala *et al.* [1988]. Further separation of conjugate pairs of defects takes place through subsequent bond-switching reactions, as illustrated in Fig. 5.7. This identification is based on the theoretical calculation of Vanderbilt and Joannopoulos [1980, 1981].

The light-induced effect on PL and mid-gap absorption has been observed in a-Chs. The thermal annealing behavior of mid-gap absorption is similar to that of the type I LESR centers, as seen in Fig. 5.2. Thus, the mid-gap absorption is associated with light-induced defects. The light-induced effect on PL appears as a luminescence fatigue, which arises from nonradiative recombination at light-induced defects.

These principal light-induced effects, that is, PD, PVC, and PDC, as mentioned above, have been simultaneously measured in a-As$_2$Se$_3$ by Nakagawa *et al.* [2010] to elucidate the correlation among them. The illumination-time dependences of $\Delta\alpha$, Δh, and

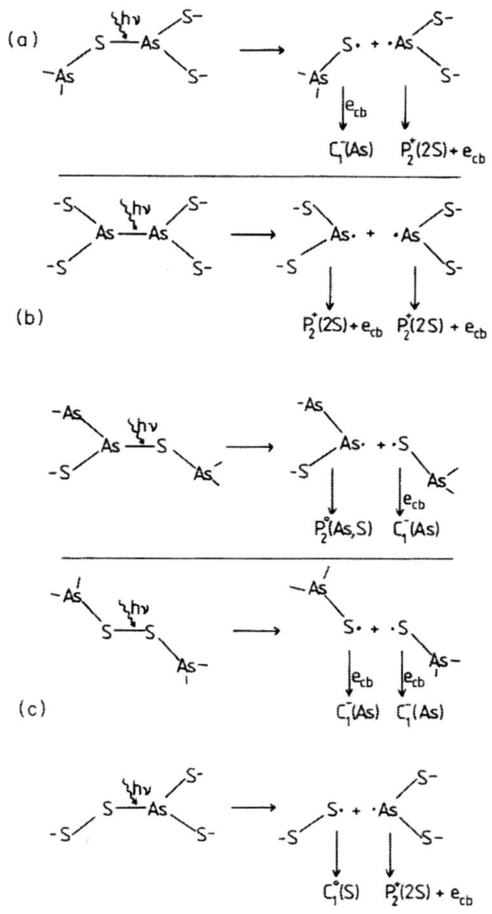

Figure 5.6   Schematic illustration of the various types of optically induced bond-breaking mechanisms operative in a-arsenic chalcogenides in (a) a chemically ordered structure, (b) the vicinity of an As–As homopolar bond, and (c) the vicinity of a S–S homopolar bond. The resulting types of metastable dangling bond defects are shown where the subscripts and superscripts to the symbols P (pnictogen) and C (chalcogen) refer, respectively, to the coordination number and charge states of the defects, with the type of nearest neighbors shown in the parentheses. Excess electrons, free to move in the conduction band, are denoted by e_{cb}. [Reproduced from Elliott and Shimakawa, *Phys. Rev. B*, **42**, 9766 (1990) by permission of American Physical Society.]

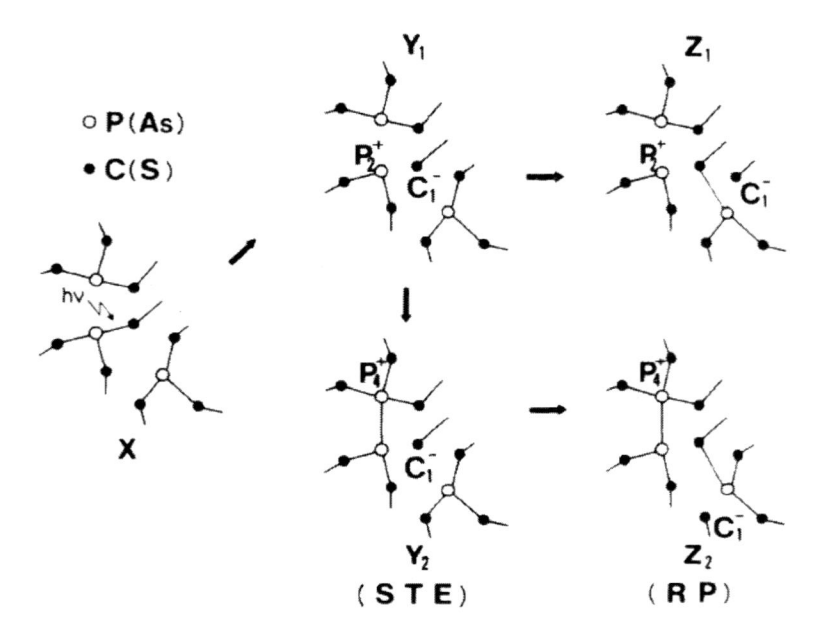

Figure 5.7 Schematic illustration of the optical generation of self-trapped exciton (STE) states (Y_1, Y_2) from the chemically ordered ground-state structure of a-As_2S_3 ($P \equiv As$, $C \equiv S$). Subsequent bond-switching reactions can lead to a greater separation between the charged defects, that is, random pairs (RP, Z_1, Z_2). [Reproduced from Shimakawa *et al.*, *Phys. Rev. B*, **42**, 11857 (1990) by permission of American Physical Society.]

$N(\tau)/N_0$ under illumination of He–Ne laser light at 633 nm and at 200 mW/cm^2 are shown in Fig. 5.8, in which $\Delta\alpha$, Δh, $N(t)$, and N_0 are the change in optical absorption coefficient, the change in surface height, the defect density at t deduced from photocurrents, and the defect density in the initial state (t = 1 s), respectively. These time evaluations are expressed by stretched exponential functions, whose β and τ are listed in Table 5.1 [Nakagawa *et al.*, 2010]. From these results, it is concluded that these light-induced effects are not correlated with each other. Nakagawa *et al.* [2010] interpreted the results in terms of the charged layer model, as shown in Fig. 5.9, as follows: The light-induced holes diffuse away to the unilluminated region so that the layers become negatively charged. A repulsive force is generated between

Table 5.1 The values of τ and β obtained for $N(t)/N_0$, Δh, and $\Delta\alpha$ [Nakagawa *et al.*, 2010]

	d = 200 nm		d = 700 nm	
	τ(s)	β	τ(s)	β
$N(t)/N_0$	18	0.55	40	0.55
Δh	70	1.0	90	1.0
$\Delta\alpha$	200	1.0	180	0.95

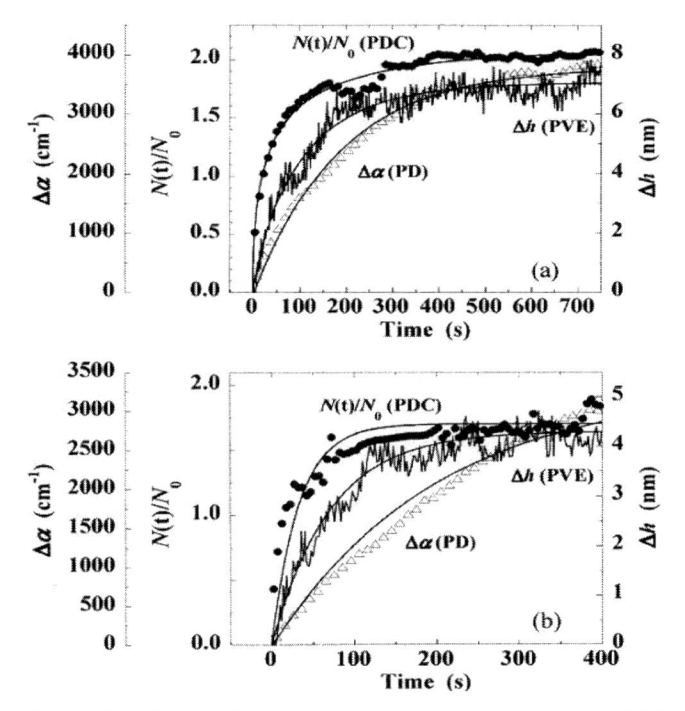

Figure 5.8 Illumination-time dependence of $\Delta\alpha$, Δh, and $N(t)/N_0$ for d (film thickness) = 700 nm, (b) d = 200 nm in a-As$_2$Se$_3$. [Reproduced from Nakagawa *et al.*, *Phys. Status Solidi C*, **7**, 857 (2010) by permission of Elsevier.]

negatively charged layers. This repulsive interlayer Coulomb interaction makes interlayer distance increased (the arrow E in Fig. 5.9). As a result, the volume expansion occurs and a slip motion along the layer takes place (the arrow S in Fig. 5.9). This lone-pair interaction between interlayers is increased, yielding the PD effect. The PDC occurs as a result of bond breaking between As and Se in Fig. 5.9 under illumination. For these processes, their principal light-induced effects occur simultaneously and independently with each other.

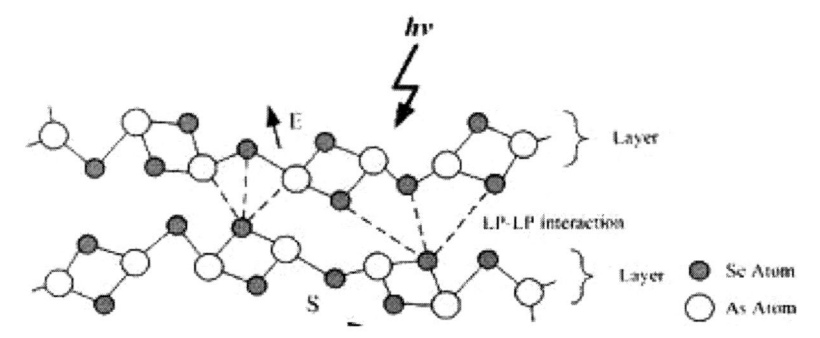

Figure 5.9   Charged layer model originally proposed for PD and PVC in a-chalcogenides. [Reproduced from Nakagawa *et al.*, *Phys. Status Solidi C*, **7**, 857 (2010) by permission of Elsevier.]

Appendix

The stretched exponential function is connected to the exponential-decay function with distribution of v through the Laplace transform [Saito and Murayama, 1987; Redfield, 1992]:

$$g(v) = \frac{1}{2\pi i} \int_{-i\infty}^{i\infty} f(t) e^{vt} dt \tag{A1}$$

$$f(t) = \exp[-(t/\tau)^\beta] \tag{A2}$$

where β and τ are a dispersion parameter and a characteristic time, respectively. $vg(v)$ is shown as a function of $v\tau$ with various values of β in Fig. A1. As seen in the figure, the small values of β correspond to a broad distribution of v, whereas the large values of β to a narrow distribution of v. For $\beta = 1$, $vg(v)$ is the δ-function, that is, $f(t)$ gives an exponential function with $v = \tau^{-1}$.

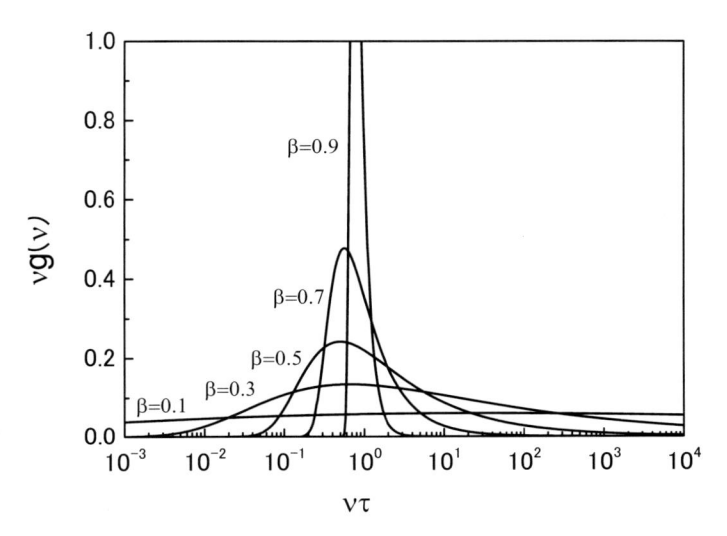

Figure A.1 *vg vs. vτ* for various values of β.

Bibliography

Amer, N. M., Skumanich, A. and Jackson, W. B. (1983). The contribution of the Staebler–Wronski effect to gap-state absorption in hydrogenated amorphous silicon, *Physica*, **117** *B*, pp. 897–898.

Aoki, T., Ikeda, K., Kobayashi, S. and Shimakawa, K. (2006). Recombination kinetics of very-long-lived photoluminescence decay in a-Si:H correlated with light-induced spin densities, *Philos. Mag. Lett.*, **86**, pp. 137–145.

Aoki, T., Komedoori, S., Kobayashi, S., Fujihashi, C., Ganjoo, A. and Shimakawa, K. (2002). Photoluminescence lifetime distribution of a-Si:H and a-Ge:H expanded to nanosecond region using wide-band frequency-resolved spectroscopy, *J. Non-Cryst. Solids*, **299–302**, pp. 642–647.

Astakov, O., Carius, R., Lambertz, A., Petrusenko, Yu., Borysenko, V., Barankov, D. and Finger, F. (2008). Structure of the ESR spectra of thin film silicon after electron bombardment, *J. Non-Cryst. Solids*, **354**, pp. 2329–2332.

Biegelsen, D. K. and Street, R. A. (1980). Photoinduced defects in chalcogenide glasses, *Phys. Rev. Lett.*, **44**, pp. 803–806.

Bishop, S. G., Strom, U. and Taylor, P. C. (1977). Optically induced metastable paramagnetic states in amorphous semiconductors, *Phys. Rev. B*, **15**, pp. 2278–2294.

Biswas, R. and Kwon, I. (1991) *New interpretations of the Staebler-Wronski effect in a-Si:H with molecular dynamics simulations*, AIP Conf. Proc. (Americal Inst. Phys. New York), pp. 45–50.

Boehme, C. and Lips, K. (2003). Theory of time-domain measurement of spin-dependent recombination with pulsed electrically detected magnetic resonance, *Phys. Rev. B*, **68**, pp. 245105-1–245105-19.

Boulitrop, F., Chenevas-Paule, A. and Dunstan, D. J. (1983). Luminescence and magnetic resonance in post-hydrogenated microcrystalline silicon, *Solid State Commun.*, **48**, pp. 181–184.

Branz, H. (1999). Hydrogen collision model: Quantitative description of metastability in amorphous silicon, *Phys. Rev. B*, **59**, pp. 5498–5512.

Branz, H. (2003). The hydrogen collision model of metastability after 5 years: experimental tests and theoretical extensions, *Sol. Energy Mater. & Sol. Cells*, **78**, pp. 425–445.

Brower, K. L. (1983). ^{29}Si hyperfine structure of unpaired spins at the Si/SiO$_2$ interface, *Appl. Phys. Lett.*, **43**, pp. 1111–1113.

Brower, K. L. (1986). Strain broadening of the dangling–bond resonance at the (111) Si-SiO$_2$ interface, *Phys. Rev. B*, **33**, pp. 4471–4478.

Butté, R., Vignoli, S., Meaudre, M., Meaudre, R., Marty, O., Savior, L. and Roca i Cabarrocas, P. (2000). Structural, optical and electronic properties of hydrogenated polymorphous silicon films deposited at 150°C, *J. Non-Cryst. Solids*, **266–269**, pp. 263–268.

Caputo, D., de Cesare, G., Irrera, F., Palma, F., Rossi, M. C., Conte, G., Nobile, G. and Fameli, G. (1994). A systematic investigation of the role of material parameters in metastability of hydrogenated amorphous silicon, *J. Non-Cryst. Solids*, **170**, pp. 278–286.

Carlson, D. E. (1986). Hydrogenated microvoids and light-induced degradation of amorphous-silicon solar cells, *Appl. Phys. A*, **41**, pp. 305–309.

Cavenett, B. C. (1981). Optically detected magnetic resonance (O. D. M. R) investigations of recombination processes in semiconductors, *Adv. Phys.*, **30**, pp. 475–538.

Cavenett, B. C., Depinna, S. P., Austin, I. G., and Searle, T. M. (1983). Determination of the g values of the ODMR signals in a-Si:H, *Philos. Mag. B*, **48**, pp. 169–185.

Chadi, D. J. and Chang, K. J. (1989). Energetics of DX-center formation in GaAs and Al$_x$Ga$_{1-x}$As alloys, *Phys. Rev. B*, **39**, pp. 10063–10074.

Cody, G. D., Tiedje, T., Abeles, B., Brooks, B. and Goldstein, Y. (1981). Disorder and the optical-absorption edge of hydrogenated amorphous silicon, *Phys. Rev. Lett.*, **47**, pp. 1480–1483.

Cohen, M. H., Fritzsche, H. and Ovshinsky, S. R. (1969). Simple band model for amorphous semiconductors alloys, *Phys. Rev. Lett.*, **22**, pp. 1065–1068.

Collins, R. W. and Paul, W. (1982). Model for the temperature dependence of photoluminescence in a-Si:H and related materials, *Phys. Rev. B*, **25**, pp. 5257–5262.

Das, D., Shirai, H., Hanna, J. and Shimizu, I. (1991). Narrow band-gap a-Si: H with improved minority carrier-transport prepared by chemical annealing, *Jpn. J. Appl. Phys.*, **30**, pp. 239–242.

de Lima, Taylor, P. C., Morrison, S, LeGeune, A. and Marques, F. C. (2002). ESR observations of paramagnetic centers in intrinsic hydrogenated microcrystalline silicon, *Phys. Rev. B*, **65**, pp. 235324-1–235324-6.

Depinna, S. P., Homewood, K., Cavenett, B. C., Austin, I. G., Searle, T. M., Willeke, G. and Kinmond, S. (1983). O.D.M.R. investigation of recombination in μc-Si:H, *Philos. Mag. B,* **47**, pp. L57–L62.

Depinna, S. P. and Dunstan, D. J. (1984). Frequency-resolved spectroscopy and its application to the analysis of recombination in semiconductors, *Philos. Mag. B,* **50**, pp. 579–597.

Dersch, H., Stuke, J. and Beichler, J. (1981). Light-induced dangling bonds in hydrogenated amorphous silicon, *Appl. Phys. Lett.,* **38**, pp. 456–458.

Dunstan, D. J. and Boulitrop, F. (1984). Photoluminescence in hydrogenated amorphous silicon, *Phys. Rev. B,* **30**, pp. 5945–5957.

Ehara, T., Ikoma, T. and Tero-Kubota, S. (2000). Multi-band electron paramagnetic resonance study of the defects in microcrystalline silicon, *J. Non-Cryst. Solids,* **266–269**, pp. 540–543.

Elliott, S. R. and Shimakawa, K. (1990). Model for bond-breaking mechanisms in amorphous arsenic chalcogenides leading to light-induced electron-spin resonance, *Phys. Rev. B,* **42**, pp. 9766–9770.

Engemann, D. and Fischer, R. (1973). Radiative recombination in amorphous silicon, *Proc. 5th Int. Conf. on Amorphous and Liquid Semiconductors,* pp. 947–952.

Engemann, D. and Fischer, R. (1974). Influence of preparation conditions on the radiative recombination in amorphous silicon, *Proc. 12th Int. Conf. on the Physics of Semiconductors,* pp. 1042–1046.

Englman, R. and Jortner, J. (1970). The energy gap law for radiative transitions in large molecules, *Mol. Phys.,* **18**, pp. 145–164.

Fehr, M., Schnegg, A., Rech, B., Lips, K., Astakov, O., Finger, F., Pfanner, G., Freysoldt, C., and Neugebauer, J., Bittl, R., and Teutloff, C. (2011). Combined multifrequency EPR and DFT study of dangling bonds in a-Si:H, *Phys. Rev. B,* **84**, pp. 245203-1–10.

Fehr, M., Schnegg, A., Teutoff, C., Bittl, R., Astakhov, O., Finger, F., Rech, B. and Lips, K. (2009). Hydrogen distribution in the vicinity of dangling bonds in hydrogenated amorphous silicon (a-Si:H), *Phys. Status Solidi A,* **207**, pp. 552–555.

Finger, F. (2004). Defects and structure of hydrogenated microcrystalline silicon films deposited by different techniques, *J. Non-Cryst. Solids,* **338–340**, pp. 168–172.

Finger, F., Carius, R., Dylla, T., Klein, S., Okur, S. and Günes, M. (2005). Instability phenomena in microcrystalline silicon films, *J. Optoelectron. Adv. Mat.,* **7**, pp. 83–90.

Finger, F., Malten, C., Hapke, P., Carius, R., Flückiger, R. and Wagner, H. (1994). Free electrons and defects in microcrystalline silicon studied by electron spin resonance, *Philos. Mag. Lett.*, **70**, pp. 247–254.

Finger, F., Müller, J., Malten, C. and Wagner, H. (1998). Electronic states in hydrogenated microcrystalline silicon, *Philos. Mag. B*, **77**, pp. 805–830.

Fontenberta, i Morral, A., Brenot, R., Hamers, E. A. G., Vanderhaghen, R. and Roca i Cabarrocas, P. (2000). In situ investigation of polymorphous silicon deposition, *J. Non-Cryst. Solids*, **266–269**, pp. 48–53.

Freysoldt, C., Pfanner, G. and Neugebauer, J. (2012). The dangling-bond defect in amorphous silicon: statistical random versus kinetically deriven defect geometries, *J. Non-Cryst. Solids*, **358**, pp. 2063–2066.

Fritzsche, H. (2001). Development in understanding and controlling the Staebler-Wronski effect in a-Si:H, *Annu. Rev. Mater. Res.*, **31**, pp. 47–79.

Fuhs, W., Mell, H., Stuke, J., Thomas, P. and Weiser, G. (1985). Photo-induced metastable effects in hydrogenated amorphous silicon (a-Si:H), *Ann. Phys.*, **42**, pp. 187–197.

Fujita,Y., Yamaguchi, M. and Morigaki, K. (1994). Kinetics of annealing of dangling bonds in sputtered amorphous silicon and hydrogenated amorphous silicon, *Philos. Mag. B*, **69**, pp. 57–67.

Ganjoo, A., Ikeda, Y. and Shimakawa, K. (2000). In situ measurements of photo-induced volume changes in amorphous chalcogenide films, *J. Non-Cryst. Solids*, **266–269**, pp. 919–923.

Gleskova, H., Morin, P. A. and Wagner, S. (1993). Kinetics of recovery of the light-induced defects in hydrogenated amorphous silicon under illumination, *Appl. Phys. Lett.*, **62**, pp. 2063–2065.

Godet, C. (1998). Metasable hydrogen atom trapping in hydrogenated amorphous silicon films: a microscopic model for metastable defect creation, *Philos. Mag. B*, **77**, pp. 765–777.

Godet, C., Morin, P. and Roca i Cabarrocas, P. (1996). Influence of the dilute-phase SiH bond concentration on the steady-state defect density in a-Si:H, *J. Non-cryst. Solids*, **198–200**, pp. 449–452.

Graeff, C. F. O., Buhleier, R. and Stutzmann, M. (1993). Light-induced annealing of metastable defects in hydrogenated amorphous silicon, *Appl. Phys. Lett.*, **62**, pp. 3001–3003.

Guéron, M. and Solomon, I. (1965). Effect of spin resonance on hot electrons by spin-orbit coupling in n-type InSb, *Phys. Rev. Lett.*, **15**, pp. 667–670.

Han, D., Yoshida, M. and Morigaki, K. (1987). Light-induced metastable defects in a-Si:H elucidated by optically detected magnetic resonance measurements at 2 K, *Solid State Commun.*, **63**, pp. 1083–1086.

Hautala, J., Ohlsen, W. D. and Taylor, P. C. (1988). Optically induced electron-spin resonance in As_xS_{1-x}, *Phys. Rev. B*, **38**, 11048–11060.

Hikita, H., Takeda, K., Kimura, Y., Yokomichi, H. and Morigaki, K. (1997). Deconvolution of ESR spectra and their light-induced effect in a-Si:H, *J. Phys. Soc. Jpn.*, **66**, pp. 1730–1740.

Hikita, H., Takeda, K., Yokomichi, H. and Morigaki, K. (unpublished).

Hirabayashi, I., Morigaki, K. and Nitta, S. (1980).New evidence for defect creation by high optical excitation in glow discharge amorphous silicon, *Jpn., J. Appl. Phys.*, **19**, pp. L357–L360.

Hirose, T., Yoshioka, H. and Horai, K. (1966). Conductivity increase of reduced rutile with the saturation of electron spin resonance, *J. Phys. Soc. Jpn.*, **21**, pp. 559–559.

Honig, A. (1966). Neutral-impurity scattering and impurity Zeeman spectroscopy in semiconductors using highly spin-polarized carriers, *Phys. Rev. Lett.*, **17**, 186–188.

Ishii, N., Kumeda, M. and Shimizu, T. (1981). The g-value of defects in amorphous C, Si and Ge. *Jpn. J. Appl. Phys.*, **20**, L673–L676.

Ishii, N. and Shimizu, T. (1998). Cluster-model calculations of hyperfine parameters and g-values fro defects in a-Si:H, *J. Non-Cryst. Solids*, **227–230**, pp. 358–361.

Isoya, J., Yamasaki, S., Okushi, H., Matsuda, A. and Tanaka, K. (1993). Electron-spin-echo envelope-modulation study of the distance between dangling bonds and hydrogen atoms in hydrogenated amorphous silicon, *Phys. Rev. B*, **47**, pp. 7013–7024.

Jackson, W. B. and Kakalios, J. (1988). Evidence for hydrogen motion in annealing of light-induced metastable defects in hydrogenated amorphous silicon, *Phys. Rev. B*, **37**, pp. 1020–1023.

Jones, R. and Lister, G. M. S. (1990). *Ab initio* calculations on metastable defects in a-Si:H. The Staebler-Wronski effect., *Philos. Mag. B*, **61**, pp. 881–894.

Kakalios, J., Street, R. A. and Jackson, W. B. (1987). Stretched-exponential relaxation arising from dispersive diffusion of hydrogen in amorphous silicon, *Phys. Rev. Lett.*, **59**, pp. 1037–1040.

Kamei, T., Hata, N., Matsuda, A., Uchiyama, T., Amano, S., Tsukamoto, K., Yoshioka, Y. and Hirao, T. (1996). Deposition and extensive light soaking of highly pure hydrogenated amorphous silicon, *Appl. Phys. Lett.*, **68**, 2380–2382.

Kamimura, H. and Mott, N. F. (1976). The variable range hopping induced by electron spin resonance in n-type silicon and germanium, *J. Phys. Soc. Jpn.*, **40**, pp. 1351–1358.

Kaplan, D., Solomon, I. and Mott, N. F. (1978). Explanation of the large spin-dependent recombination effect in semiconductors, *J. de Physique*, **39**, pp. L51–L54.

Kastner, M., Adler, D. and Fritzsche, H. (1976). Valence-alternation model for localized gap states in lone-pair semiconductors, *Phys. Rev. Lett.*, **37**, pp. 1504–1507.

Kimerling, L. C. (1978). Recombination enhanced defect reactions, *Solid State Electron.*, **21**, pp. 1391–1401.

Kishimoto, N. and Morigaki, K. (1977). Resistivity decrease due to electron spin resonance in the metallic region of heavily phosphorus-doped silicon, *J. Phys. Soc. Jpn.*, **42**, pp. 137–145.

Kishimoto, N., Morigaki, K. and Murakami, K. (1981). Conductivity change due to electron spin resonance in amorphous Si-Au system, *J. Phys. Soc. Jpn.*, **50**, pp. 1970–1977.

Kondo, M. and Morigaki, K. (1990). Light-induced phenomena in a-Si:H as elucidated by optically detected electron nuclear double resonance, *Proc. Int. Conf. Physics of Semiconductors,* eds. Anastassakis, E. M. and Joannopoulos, J. D. (World Scientific, Singapore), pp. 2083–2086.

Kondo, M and Morigaki, K. (1991). The role of hydrogen clusters in the Staebler-Wronski effect of amorphous silicon as elucidated by optically detected electron nuclear double resonance, *J. Non-Cryst. Solids*, **137–138**, pp. 247–250.

Kondo, M. and Morigaki, K. (1993). Possibility of hydrogen migration in photoinduced defect creation process of a-Si:H, *J. Non-Cryst. Solids*, **164–166,** pp. 227–230.

Kondo, M., Nishimiya, T., Saito, K. and Matsuda, A. (1998). Light induced phenomena in microcrystalline silicon, *J. Non-Cryst. Solids*, **227–230**, pp. 1031–1035.

Kondo, M., Yamasaki, S. and Matsuda, A. (2000). Microscopic structure of defects in microcrystalline silicon, *J. Non-Cryst. Solids*, **266–269**, pp. 544–547.

Kugler, S. (2012). Advances in understanding the defects contributing to the tail states in pure amorphous silicon, *J. Non-Cryst. Solids*, **358**, pp. 2060–2062.

Lang, D. V. (1992). *Deep Centers in Semiconductors: A State-of-Art Approach*, S. T. Chapter 7, "DX Centers in III–V Alloys" (Gordon and Breach Science Publishers, Switzerland), pp. 591–642.

Lang, D. V. and Kimerling, L. C. (1974). Observation of recombination-enhanced defect reactions in semiconductors, *Phys. Rev. Lett.*, **33**, pp. 489–492.

Lang, D. V. and Kimerling, L. C. (1976). Observation of athermal defect annealing in GaP, *Appl. Phys. Lett.*, **28**, pp. 248–250.

Lang, D. V. and Logan, R. A. (1977). Large-lattice-relaxation model for persistent photoconductivity in compound semiconductors, *Phys. Rev. Lett.*, **39**, pp. 635–639.

Lang, D. V. and Logan, R. A. (1979). *Physics of Semiconductors*, 1978, Conference Series No. 43, ed. Wilson, B. L. H., Chapter 13 "Defects Chemical Shifts of DX Centers in $Al_xGa_{1-x}As$" (Institute of Physics, Bristol and London), pp. 433–436.

Lepine, D. (1972). Spin-dependent recombination on silicon surface, *Phys. Rev. B*, **6**, pp. 436–441.

Lepine, D., Grazhulis, V. A., and Kaplan, D. (1976). Spin-dependent recombination on dislocation in silicon, *Proc. 13th Int. Conf. Phys. Semicond.*, Rome, ed. Fumi, E. G., pp. 1081–1084.

Lips, K., Boehme, C. and Ehara, T. (2005). The impact of the electron spin on charge carrier recombination: the example of amorphous silicon, *J. Optoelec. Adv. Mat.*, **7**, pp. 13–24.

Lips, K., Kanschat, P. and Fuhs, W. (2003). Defects and recombination in microcrystalline silicon, *Sol. Energy Mat. & Sol. Cells*, **78**, pp. 513–541.

Longeaud, C., Roy, D. and Saadane, O. (2002). Role of interstitial hydrogen and voids in light-induced metastable defect formation in hydrogenated amorphous silicon: A model, *Phys. Rev. B*, **65**, pp. 085206-1–085206-9.

Lukács, R. and Kugler, S. (2011). Photoinduced volume changes of obliquely and flatly deposited amorphous AsSe films universal description of the kinetics, *Jpn. J. Appl. Phys.*, **50**, pp. 091401-1–091401-4.

Lukács, R., Veres, M., Shimakawa, K. and Kugler, S. (2010). On photoinduced volume change in amorphous selenium: Quantum chemical calculation and Raman spectroscopy, *J. Appl. Phys.*, **107**, pp. 073517-1–073517-5.

Malten, C., Carius, R., Finger, F. and Yamasaki, S. (1998). ODMR measurements of microcrystalline silicon, *MRS Proc.*, **507**, pp. 787–792.

Maxwell, R. and Honig, A. (1966). Neutral-impurity scattering experiments in silicon with highly spin-polarized electrons, *Phys. Rev. Lett.*, **17**, pp. 188–190.

Matsuda, A. (2006) *Springer Handbook of Electronic and Photonic Materials*, eds. Kasap, S. and Capper, P., Part C 26 "morphous and Microcrystalline Silicon" (Springer, New York), pp. 581–595.

Mooney, P. M. (1992). *Deep Centers in Semiconductors: A State-of-Art Approach*, ed. Pantelides, S. T., Chapter 7, "DX Centers in III–V Alloys:

Recent Developments" (Gordon and Breach Science Publishers, Switzerland), pp. 643–665.

Morigaki, K. (1981). Spin-dependent radiative and nonradiative recombinations in hydrogenated amorphous silicon: Optically detected magnetic resonance, *J. Phys. Soc. Jpn.,* **50**, pp. 2279–2287.

Morigaki, K. (1983). Optically detected magnetic resonance in amorphous semiconductors, *Jpn. J. Appl. Phys.,* **22**, pp. 375–388.

Morigaki, K. (1984) *Semiconductors and Semimetals 21 Part C,* ed. Pankove, J. I., Chapter 4 "Optically Detected Magnetic Resonance" (Academic Press, Orlando), pp. 155–191.

Morigaki, K. (1988). Microscopic mechanism for the photo-creation of dangling bonds in a-Si:H, *Jpn. J. Appl. Phys.,* **27**, pp. 163–168.

Morigaki, K. (1992). Self-trapping of holes and light- and thermally induced defect creation in a-Si:H, *J. Non-Cryst. Solids,* **141**, pp. 166–175.

Morigaki, K. (1999) *Physics of Amorphous Semiconductors* (World Scientific, Singapore, and Imperial College Press, London).

Morigaki, K. (2001). Role of hydrogen-related weak bonds and hydrogen-related dangling bonds in light-induced structural changes in a-Si:H, *Bull. Hiroshima Inst. Tech,* **35**, pp. 47–56.

Morigaki, K. (2002a). Photoluminescence in amorphous solids, *Nonlinear Opt.,* **29**, pp. 265–272.

Morigaki, K. (2002b). On the electron spin resonance of band tails and defects in hydrogenated amorphous silicon, *Bull. Hiroshima Inst. Tech.,* **36**, pp. 25–30.

Morigaki, K. (2009). Recombination processes and light-induced defect creation in hydrogenated amorphous silicon, *Phys. Status Solidi A,* **206**, pp. 868–873.

Morigaki, K. and Hikita, H. (1999). Hydrogen-related dangling bonds in hydrogenated amorphous silicon, *Bull. Hiroshima Inst. Tech.,* **33**, pp. 135–142.

Morigaki, K. and Hikita, H. (2005). Light-induced creation of hydrogen-pairs in a-Si:H, *Solid State Commun.,* **136**, pp. 616–620.

Morigaki, K. and Hikita, H. (2007). Modeling of light-induced defect creation in hydrogenated amorphous silicon, *Phys. Rev. B,* **76**, pp. 085201-17.

Morigaki, K. and Hikita, H. (2011). Stretched exponential relaxation processes in hydrogenated amorphous and polymorphous silicon, *Phys. Status Solid C,* **8**, pp. 2564–2568.

Morigaki, K. and Hikita, H. (unpublished).

Morigaki, K., Hikita, H. and Kondo, M. (1995). Self-trapping of holes and related phenomena in a-Si:H, *J. Non-Cryst. Solids*, **190**, pp. 38–47.

Morigaki, K., Hikita, H. and Ogihara, C. (2009a). Light-induced defect creation under intense pulsed illumination in hydrogenated amorphous silicon, *J. Optoelec. Adv. Mater.*, **11**, pp. 1–14.

Morigaki, K., Hikita, H., Takeda, K. and Roca I Cabarrocas, P. (2010). The kinetics of the light-induced defect creation in hydrogenated polymorphous silicon–stretched exponential relaxation, *Phys. Status Solidi C*, **7**, pp. 692–695.

Morigaki, K., Hikita, H., Takemura, H., Yoshimura, T. and Ogihara, C. (2003). Light-induced defect creation under pulsed subbandgap illumination in hydrogenated amorphous silicon, *Philos. Mag. Lett.*, **83**, pp. 341–349.

Morigaki, K., Hikita, H., Yamaguchi, M. and Fujita, Y. (1998). The structure of dangling bonds having hydrogen at a nearby site in a-Si:H, *J. Non-Cryst. Solids*, **227–230**, pp. 338–342.

Morigaki, K., Hikita, H., Yamaguchi, M. and Fujita, Y. (2003). Anisotropic magnetic centers and conduction electrons in hydrogenated micro-crystalline silicon, *Mat. Eng.*, *B*, **103**, pp. 37–44.

Morigaki, K., Hirabayashi, I., Nakayama, M., Nitta, S. and Shimakawa, K. (1980). Fatigue effect in luminescence of glow discharge amorphous silicon at low temperatures, *Solid State Commun.*, **33**, pp. 851–856.

Morigaki, K., Kishimoto, N. and Lepine, D. J. (1975). Resistivity decreases due to electron spin resonance in the metallic region of heavily phosphorus doped silicon, *Solid State Commun.*, **17**, pp. 1017–1019.

Morigaki, K. and Kondo, M. (1995). Optically detected magnetic resonance in a-Si:H, *Solid State Phenomena*, **44–46**, pp. 731–764.

Morigaki, K. and Niikura, C. (2005). The *g*-values of defects in hydrogenated microcrystalline silicon, *Solid State Commn.*, **136**, pp. 308–312.

Morigaki, K., Niikura, C., Bourée, J. E. and Equer, B. (2002). Anisotropic magnetic centres in hydrogenated microcrystalline and polymorphous silicon, *J. Non-Cryst. Solids*, **299–302**, pp. 561–565.

Morigaki, K., Niikura, C., Hikita, H. and Yamaguchi, M. (2009b). Defects in hydrogenated microcrystalline silicon prepared by plasma-enhanced chemical vapour deposition, *J. Appl. Phys.*, **105**, 083703-1–083703-6.

Morigaki, K., Niikura, C. and Hikita, H. (2009c). Defect states in hydro-genated microcrystalline silicon, *Kotai Butsuri* (Solid State Physics), **44**, pp. 211–225 (in Japanese).

Morigaki, K. and Ogihara, C. (2006) *Springer Handbook of Electronic and Photonic Materials*, eds. Kasap, S. and Capper, P., "Structural, Optical, and Electronic Properties"(Springer, New York), pp. 565–580.

Morigaki, K. and Onda, M. (1972). Resistivity decrease due to donor spin resonance in n-type germanium, *J. Phys. Soc. Jpn.,* **33**, pp. 1031–1046.

Morigaki, K., Sano, Y. and Hirabayashi, I. (1982). Defect creation by optical excitation in hydrogenated amorphous silicon as elucidated by optically detected magnetic resonance, *J. Phys. Soc. Jpn.,* **51**, pp. 147–152.

Morigaki, K., Takeda, K., Hikita, H., Ogihara, C. and Roca i Cabarrocas, P. (2008). The kinetics of the light-induced defect creation in hydrogenated amorphous silicon—stretched exponential relaxation-, *J. Non-Cryst. Solids*, **354**, pp. 2131–2134.

Morigaki, K., Takeda, K., Hikita, H. and Roca i Cabarrocas, P. (2005a). Light-induced defect creation in hydrogenated polymorphous silicon, *Mat. Sci. Eng. B*, **121**, pp. 34–41.

Morigaki, K., Takeda, K., Hikita, H. and Roca i Cabarrocas, P. (2005b). Light-induced defect creation in hydrogenated polymorphous silicon during repeated cycles of illumination and annealing, *Philos. Mag.*, **85**, pp. 3393–3407.

Morigaki, K., Takeda, K., Hikita, H. and Roca i Cabarrocas, P. (2007). Dispersive processes of light-induced defect creation in hydrogenated amorphous silicon, *Solid State Commun.*, **142**, pp. 232–236.

Morigaki, K. and Toyotomi, S. (1971). Spin energy transfer from donor spin system to mobile electron system in P-doped Si, *J. Phys. Soc. Jpn.,* **30**, pp. 1207–1208.

Morigaki, K., Yamaguchi, M., Hirabayashi, I. and Hayasi, R. (1987). *Disordered Semiconductors*, ed. Kastner, M. A., Thomas, G. A. and Ovshinsky, S. R., "Defects in a-Si:H" (Plenum Press, New York) pp. 415–424.

Morigaki, K. and Yoshida, M. (1985). Gap states in hydrogenated amorphous silicon. The origins of the light-induced ESR and the low-energy luminescence, *Philos. Mag. B*, **52**, pp. 289–298.

Mott, N. F. (1970). Conduction in non-crystalline systems IV. Anderson localization in a disordered lattices, *Philos. Mag.*, **22**, pp. 7–29.

Mott, N. F. (1978). Recombination: A survey, *Solid State Electron.*, **21**, pp. 1275–1280.

Mott, N. F. and Davis, E. A. (1979) *Electronic Processes in Non-Crystalline Materials*, 2nd Ed. (Clarendon Press, Oxford).

Movaghar, B., Ries, B. and Schweitzer, L. (1980). Theory of the resonant and non-resonant luminescence changes in amorphous silicon, *Philos. Mag. B*, **41**, pp. 141–157.

Müller, J., Finger, F., Carius, R. and Wagner, H. (1999). Electron spin resonance investigation of electronic states in hydrogenated microcrystalline silicon, *Phys. Rev. B*, **60**, pp. 11666–11677.

Murayama, K., Morigaki, K. and Kanzaki, H. (1975). Exchange effects in optical detection ESR and dynamics of the optical pumping cycle of F-centers in KCl, *J. Phys. Soc. Jpn.*, **38**, pp. 1623–1633.

Nakagawa, N., Shimakawa, K., Itoh, T. and Ikeda, Y. (2010). Dynamics of principal photoinduced effects in amorphous chalcogenides: In-situ simultaneous measurements of photodarkening, volume changes, and defect creation, *Phys. Status Solidi C*, **7**, pp. 857–860.

Nakamura, N., Takahama, T., Isomura, M., Nishikuni, M.,Tarui, H., Wakisaka, K., Tsuda, S., Nakano, S., Kishi, Y. and Kuwano, Y. (1992). Irradiation-temperature dependence of the light-induced effect in a-Si solar cells, *Jpn. J. Appl. Phys.*, **31**, pp. 1267–1271.

Nelson, R. J. (1977). Long-lifetime photoconductivity effect in n-type GaAlAs, *Appl. Phys. Lett.*, **31**, pp. 351–353.

Neto, A. L. B., Dylla, T., Klein, S., Repmann, T., Lambertz, A., Carius, R. and Finger, F. (2004). Defects and structure of hydrogenated microcrystalline silicon films deposited by different techniques, *J. Non-Cryst. Solids*, **338–340**, pp. 168–172.

Nickel, N. H., Jackson, W. B. and Johnson, N. M. (1993). Light-induced creation of metastable paramagnetic defects in hydrogenated polycrystalline silicon, *Phys. Rev. Lett.*, **71**, pp. 2733–2736.

Nielsen, B. Bech, Johannesen, P., Stallinga, P. and Nielsen, K. Bonde (1997). Identification of the silicon vacancy containing a single hydrogen atom by EPR, *Phys. Rev. Lett.*, **79**, pp. 1507–1510.

Niikura, C., Roca I Cabarrocas, P. and Bourée, J-E. (2011). Structural properties of microcrystalline Si films prepared by hot-wire/catalytic chemical vapor deposition under conditions close to the transition from amorphous to microcrystalline growth, *Thin Solid Films*, **519**, pp. 4502–4505.

Nishi, Y. (1971). Study of silicon-silicon dioxide structure by electron spin resonance I, *Jpn. J. Appl. Phys.*, **10**, pp. 52–62.

Nonomura, S., Yoshida, N., Gotoh, T., Sakamoto, T., Kondo, M., Matsuda, A. and Nitta, S. (2000). The light-induced metastable lattice expansion in hydrogenated amorphous silicon, *J. Non-Cryst. Solids*, **266–269**, pp. 474–480.

Ogihara, C. (1998). Frequency resolved spectroscopy in a-Si:H/a-Si$_{1-x}$N$_x$:H multilayers and band-edge modulated a-Si$_{1-x}$N$_x$:H alloys, *J. Non-Cryst. Solids*, **227–230**, pp. 517–522.

Ogihara, C., Inagaki, Y. and Morigaki, K. (2011). Light-induced effects on the radiative recombination rate of electron-hole pairs in a-Si:H, *Phys. Status Solidi C*, **8**, pp. 2792–2795.

Ogihara, C., Inagaki, Y., Taketa, A. and Morigaki, K. (2012). Thermal quenching of defect photoluminescence and recombination rates of electron-hole pairs in a-Si:H, *J. Non-Cryst. Solids*, **358**, pp. 2004–2006.

Ogihara, C., Nomiyama, T., Yamamoto, H., Nakanishi, K., Harada, J., Yu, X. and Morigaki, K. (2006). Light-induced creation of defects related to low energy photoluminescence in hydrogenated amorphous silicon, *J. Non-Cryst. Solids*, **352**, pp. 1064–1067.

Ogihara, C., Takemura, H., Yoshida, H. and Morigaki, K. (2000). Lifetime distribution of PL under pulsed excitation in hydrogenated amorphous silicon based films, *J. Non-Cryst. Solids*, **266–269**, pp. 574–577.

Ogihara, C., Yu, X. and Morigaki, K. (2008). Analysis of lifetime distribution of defect luminescence in hydrogenated amorphous silicon, *J. Non-Cryst. Solids*, **354**, pp. 2121–2125.

Oheda, H. (1999). Real-time modulation of Si-H vibration in hydrogenated amorphous silicon, *Phys. Rev. B*, **60**, pp. 16531–16542.

Palinginis, K. C., Cohen, J. C., Guha, S. and Yang, J. C. (2001). Experimental evidence indicating a global mechanism for light-induced degradation in hydrogenated amorphous silicon, *Phys. Rev. B*, **63**, pp. 201203-1-201203-4.

Pankove, J. I. and Berkeyheiser, J. E. (1980). Light-induced radiative recombination centers in hydrogenated amorphous silicon, *Appl. Phys. Lett.*, **37**, pp. 705–706.

Pantelides, S. T., Chapter 7 "DX Centers in III–V Alloys" (Gordon and Breach Science Publishers, Switzerland), pp. 591–541.

Pfanner, G., Freysoldt, C. and Neugebauer, J. (2011). *Ab initio* study of electron paramagnetic resonance hyperfine structure of the silicon dangling bond: Role of the local environment, *Phys. Rev. B*, **83**, pp. 144110-1-144110-8.

Poissant, Y., St'ahel, P. and Roca i Cabarrocas, P. (2000). Effects of temperature on the kinetics of metastable defect creation in polymorphous and amorphous silicon materials and solar cells, *Proc. 16th European Photovoltaic and Solar energy conference*, pp. 377–380.

Redfield, D. (1986). Reversibility of recombination-induced defect reactions in amorphous Si:H, *Appl. Phys. Lett.*, **49**, pp. 1517–1518.

Redfield, D. (1992). Interpretation of stretched-exponential defect kinetics in a-Si:H, *MRS Proc.*, **258**, pp. 341–346.

Redfield, D. and Bube, R. H. (1990). Identification of defects in amorphous silicon, *Phys. Rev. Lett.*, **65**, pp. 464–467.

Redfield, D. and Bube, R. H. (1996) *Photoinduced Defects in Semiconductors* (Cambridge University Press, UK).

Roca i Cabarrocas, P. (2000). Plasma enhanced chemical vapor deposition of amorphous, polymorphous and microcrystalline silicon films, *J. Non-Cryst. Solids*, 266–269, pp. 31–37.

Roca i Cabarrocas, P., Châbane, N., Kharchenko, A. V. and Tchakarov, S. (2004). Polymorphous silicon thin films produced in dusty plasmas: application to solar cells, *Plasma Phys. Control. Fussion*, **46**, pp. B235–B243.

Roca i Cabarrocas, P., Fontcuberta i Morral, A. and Possant, Y. (2002). Growth and optoelectronic properties of polymorphous silicon thin films, *Thin Solid Films*, **403–404**, pp. 39–46.

Roca i Cabarrocas, P., Hamma, S., Sharma, S. N., Viera, G., Bertran, E. and Costa, J. (1989). Nanoparticle formation in low-pressure silane plasmas: bridging the gap between a-Si:H and μc-Si films, *J. Non-Cryst. Solids*, **227–230**, pp. 871–875.

Roy, D., Longeaud, C., Saadane, O., Gueunier, M. E., Vignoli, S., Butté, R., Meaudre, R. and Meaudre, M. (2002). Evolution with light-soaking of polymorphous material prepared at 423 K, *J. Non-Cryst. Solids*, **299–302**, pp. 482–486.

Saito, R. and Murayama, K. (1987), A universal distribution function of relaxation in amorphous materials, *Solid State Commun.*, **63**, pp. 625–627.

Saleh, Z. M., Tarui, H., Ninomiya, K., Takahama, T., Nakashima, Y., Nakamura, N., Haku, H., Wakisaka, K., Tanaka, M., Tsuda, S., Nakano, S., Kishi, M. and Kuwano, Y. (1992). Transient light-induced ESR investigations of the role of hydrogen in the stability of a-Si:H, *Jpn. J. Appl. Phys.*, **31**, pp. 995–998.

Saleh, Z. M., Tarui, H., Takahama, T., Nakamura, N., Nishikuni, M., Tsuda, S., Nakano, S. and Kuwano, Y. (1993). Trapping and recombination of photogenerated carriers in as-grown high-temperature-annealed and light-soaked a-Si:H, *Jpn. J. Appl. Phys.*, **32**, pp. 3376–3384.

Schmidt, J. and Solomon, I. (1966). Modulation de la photoconductivité dans le silicium á basse temperature par resonance magnétique électronique des impurities peu profondes, *Compt. Rend.*, **263**, pp. 169–172.

Schultz, N. and Taylor, P. C. (1999). Low temperature kinetics for the growth and decay of band-tail carriers and dangling bonds in hydrogenated amorphous silicon, *MRS Proc.*, **557**, pp. 353–358.

Schultz, N., Vardeny, Z. V. and Taylor, P. C. (1997). *Spin dependent photoinduced absorption in a-Si:H, MRS Proc.*, Vol. 467, eds. Wagner, S., Hack, M., Schiff, E. A., Schropp, R. and Shimizu, I. (*MRS Proc.*, Pittsburgh), pp. 179–183.

Schumm, G., Jackson, W. B. and Street, R. A. (1993). Nonequilibrium occupancy of tail states and defects in a-Si:H: Implication for defect structure, *Phys. Rev. B*, **48**, pp. 14198–14207.

Senda, M., Yoshida, N. and Shimakawa, K. (1999). Kinetics of photoinduced defect creation in amorphous semiconductors: analogy to a logistic equation in a biological system, *Philos. Mag. Lett.*, **79**, pp. 375–381.

Shimakawa, K. (1985). Exciton recombination in amorphous chalcogenides, *Phys. Rev. B*, **31**, pp. 4012–4014.

Shimakawa, K. and Elliott, S. R. (1988). Reversible photoinduced change of ac conduction in amorphous As_2S_3 films, *Phys. Rev. B*, **38**, pp. 12479–12482.

Shimakawa, K., Ikeda, Y. and Kugler, S. (2004). *Non-Crystalline Materials for Optoelectronics* Vol. 1, ed. Popescu, M., Chapter 5 "Fundamental Optoelectronic Processes in Amorphous Chalcogenides" (INOE Publishing House, Bucharest), pp. 103–130.

Shimakawa, K., Inami, S. and Elliott, S. R. (1990). Reversible photoinduced change of photoconductivity in amorphous chalcogenide films, *Phys. Rev. B*, **42**, pp. 11857–11861.

Shimakawa, K., Kolobov, A. V. and Elliott, S. R. (1995). Photoinduced effects and metastability in amorphous semiconductors and insulators, *Adv. Phys.*, **44**, pp. 475–588.

Shimakawa, K., Meherun-Nessa, Ishida, H. and Ganjoo, A. (2004). Quantum efficiency of light-induced defect creation in hydrogenated amorphous silicon and amorphous As_2Se_3, *Philos. Mag. Lett.*, **84**, pp. 81–89.

Shimizu, T. (2004). Stabler-Wronski effect in hydrogenated amorphous silicon and related alloy films, *Jpn. J. Appl. Phys.*, **43**, pp. 3257–3268.

Shinozuka, Y. and Toyozawa, Y. (1979). Self-trapping in mixed crystal-Clustering, dimensionality, percolation, *J. Phys. Soc. Jpn.*, **46**, pp. 505–514.

Singh, J. and Shimakawa, K. (2003) *Advances in Amorphous Semiconductors* (Taylor & Francis, London).

Solomon, I. (1972) *Spin-dependent transport in semiconductors*, Proc. 11th International Conf. Phys. Semiconductors, Vol. 1 (Polish Sci. Pub., Warsow), pp. 27–37.

Solomon, I. (1976). Spin-dependent recombination in a silicon p–n junction, *Solid State Commun.*, **20**, pp. 215–217.

Solomon, I. (1979) *Amorphous Semiconductors*, ed. Brodsky, M. H., Chapter 7, "Spin effects in Amorphous Semiconductors" (Springer, Berlin), pp. 189–213.

Solomon, I. and Bhatnagar, M. (1992). Transport dependant du spin dans une diode cristalline, *Compt. Rend*, **314**, pp. 1133–1138.

Solomon, I., Biegelsen, D. and Knights, J. C. (1977). Spin-dependent photoconductivity in n-type and p-type amorphous silicon, *Solid State Commun.*, **22**, pp. 505–508.

Stachowitz, R., Schubert, M. and Fuhs, W. (1994). Frequency-resolved spectroscopy and its application to low-temperature geminate recombination in a-Si:H, *Philos. Mag. B*, **70**, pp. 1219–1230.

Staebler, D. L. and Wronski, C. R. (1977). Reversible conductivity changes in discharge-produced amorphous Si, *Appl. Phys. Lett.*, **31**, pp. 292–294.

Staebler, D. L. and Wronski, C. R. (1980). Optically induced conductivity changes in discharge-produced hydrogenated amorphous silicon, *J. Appl. Phys.*, **51**, pp. 3262–3268.

Stradins, P. and Fritzsche, H. (1996). Light-induced metastable changes in defect density and photoconductivity of a-Si:H between 4.2 and 300 K, *J. Non-Cryst. Solids*, **198–200**, pp. 432–435.

Stradins, P., Kondo, M. and Matsuda, A. (2000). A study of light-induced degradation of a-Si:H by nanosecond laser pulse pairs of variable delay, *J. Non-Cryst. Solids*, **266–269**, pp. 405–409.

Street, R. A. (1991). *Hydrogenated Amorphous Silicon* (Cambridge University Press, UK).

Street, R. A. and Biegelsen, D. K. (1980). Luminescence and ESR studies of defects in hydrogenated amorphous silicon, *Solid State Commun.*, **33**, pp. 1159–1162.

Street, R. A., Knights, J. C. and Biegelsen, D. K. (1978). Luminescence studies of plasma-deposited hydrogenated silicon, *Phys. Rev. B*, **18**, pp. 1880–1891.

Street, R. A. and Mott, N. F. (1975). States in the gap in glassy semiconductors, *Phys. Rev. Lett.*, **35**, pp. 1293–1296.

Stutzmann, M. and Biegelsen, D. K. (1983). Electron-spin-lattice relaxation in amorphous silicon and germanium, *Phys. Rev. B*, **28**, pp. 6256–6261.

Stutzmann, M. and Biegelsen, D. K. (1986). Electron-nuclear double-resonance experiments in hydrogenated amorphous silicon, *Phys. Rev. B*, **34**, pp. 3093–3107.

Stutzmann, M., Brandt, M. S. and Bayerl, M. W. (2000). Spin-dependent processes in amorphous and microcrystalline silicon: a survey, *J. Non-Cryst. Solids*, **266–269**, pp. 1–22.

Stutzmann, M., Jackson, W. B. and Tsai, C. C. (1985). Light-induced metastable defects in hydrogenated amorphous silicon: A systematic study, *Phys. Rev. B*, **32**, pp. 23–47.

Stutzmann, M., Rossi, M. C. and Brandt, M. S. (1994). Pulsed-light soaking of hydrogenated amorphous silicon, *Phys. Rev. B*, **50**, pp. 11592–11605.

Su, T., Ju, T., Yan, B., Yang, J., Guha, S. and Taylor, P. C. (2008). ESR study of the hydrogenated nanocrystalline silicon thin films, *J. Non-Cryst. Solids*, **354**, pp. 2231–2234.

Su, T., Taylor, P. C., Ganguly, G. and Carlson, D. E. (2002). Direct role of hydrogen in the Staebler-Wronski effect in hydrogenated amorphous silicon, *Phys. Rev. Lett.*, **89**, pp. 015502-1–015502-4.

Su, T., Taylor, P. C., Ganguly, G. and Carlson, D. E. (2004). A hydrogen-related defect and the Staebler-Wronski effect in hydrogenated amorphous silicon, *J. Non-Cryst. Solids*, **338–340**, pp. 357–360.

Sumi, H. (1984). Dynamic defect reaction induced by multiphonon nonradiative recombination of injected carriers at deep levels in semiconductors, *Phys. Rev. B*, **29**, pp. 4616–4630.

Sumi, H. (1988). Nonradiative electron–hole recombination at deep-level defects and recombination enhanced defect reactions in semiconductors, *Kotai Butsuri* (Solid State Physics), **23**, pp. 221–230 (in Japanese).

Takeda, K., Hikita, H., Kimura, Y., Yokomichi, H., Yamaguchi, M. and Morigaki, K. (1997). Light-induced annealing f dangling bonds in a-Si:H, *Jpn. J. Appl. Phys.*, **36**, pp. 991–996.

Takeda, K., Hikita, H., Kimura, Y., Yokomichi, H. and Morigaki, K. (1998). Electron spin resonance study of light-induced annealing of dangling bonds in glow discharge hydrogenated amorphous silicon: Deconvolution of electron spin resonance spectra, *Jpn. J. Appl. Phys.*, **37**, pp. 6309–6317.

Takeda, K., Hikita, H., Kimura, Y., Yokomichi, H., Izumi, T. and Morigaki, K. (2000). Electron spin resonance spectra and their thermal annealing effects in hydrogenated amorphous silicon, *Proceedings of the School of Engineering, Tokai Univ.*, Series *E*, **25**, 17–25.

Takeda, K., Morigaki, K., Hikita, H. and Roca i Cabarrocas P. (2008). Thermal annealing effects of dangling bonds in hydrogenated polymorphous silicon, *J. Appl. Phys.*, **104**, pp. 053715-1-053715-6.

Takemura, H., Ogihara, C. and Morigaki, K. (2002). Lifetime resolved study of photoluminescence under pulsed excitation in a-Si:H based films, *J. Phys. Soc. Jpn.*, **71**, pp. 625–629.

Takenaka, H., Ogihara, C. and Morigki, K. (1988). Time-resolved optically detected magnetic resonance experiment on conduction band tail electrons and A centre in hydrogenated amorphous silicon, *J. Phys. Soc. Jpn.*, **57**, pp. 3858–3867.

Tanaka, Ke. and Shimakawa, K. (2011) *Amorphous Chalcogenide Semiconductors and Related Materials* (Springer, New York).

Taylor, P. C. (2006). The localization of electrons in amorphous semi-conductors: A twenty-first century perspective, *J. Non-Cryst. Solids*, **352**, pp. 839–850.

Tiedje, T. (1984) *Semiconductors and Semimetals 21 Part C*, ed. Pankove, J. I., Chapter 6 "Information about band-tail states from time-of-flight experiments" (Academic Press, Orlando), pp. 207–238.

Timilsina, R. and Biswas, P. (2010). Theoretical study of hydrogen microstructure in models of hydrogenated amorphous silicon, *Phys. Status Solidi A*, **207**, pp. 609–612.

Toyoda, Y. and Hayashi, Y. (1970). Bolometric detection of ESR in P-doped Si at low temperature, *J. Phys. Soc. Jpn.*, **29**, pp. 247–248.

Toyoda, Y. and Hayashi, Y. (1971). The spin dependent conductivity in P-doped Si, *J. Phys. Soc. Jpn.*, **30**, 1511–1512.

Toyotomi, S. (1974). Resistance change induced by electron spin resonance and spin relaxation in phosphorus-doped silicon, *J. Phys. Soc. Jpn.*, **37**, pp. 130–139.

Toyotomi, S. and Morigaki, K. (1970). Microwave hot electron effect and resistivity decrease due to donor spin resonance in P-doped Si, *Solid State Commun.*, **8**, pp. 1307–1308.

Toyozawa, Y. (2003) *Optical Processes in Solids* (Cambridge University Press, UK).

Tsang, T. and Street, R. A. (1979). Recombination in plasma-deposited amorphous Si:H. Luminescence decay, *Phys Rev. B,* **19**, pp. 3027–3040.

Tzanetakis, P., Kopidakis, N., Androulidaki, M., Kalouzos, C., Stradins, P. and Fritzsche, H. (1996). Short-laser-pulse and steady-light induced degradation of intrinsic, p-type and compensated a-Si:H, *J. Non-Cryst. Solids*, **198–200**, pp. 458–461.

Uchino, T., Takahashi, M. and Yoko, T. (2001). E' centers in amorphous SiO_2 revisited: A new look at an old problem, *Phys. Rev. Lett.*, **86**, pp. 5522–5525.

Uchino, T. and Yoko, T. (2003). Mechanism of electron localization at edge-sharing units in amorphous SiO_2, *Phys. Rev. B*, **68**, pp. 041201- –041201-4.

Umeda, T., Yamasaki, S., Isoya, J., Matsuda, A. and Tanaka, K. (1996). Electronic structure of band-tail electrons in a-Si:H, *Phys. Rev. Lett.*, **77**, pp. 4600–4603.

Vanderbilt, D. and Joannopoulos, J. D. (1980). Theory of defect states in glassy selenium, *Phys. Rev. B*, **22**, pp. 2927–2939.

Vanderbilt, D. and Joannopoulos, J. D. (1981). Theory of defect states in glassy As_2Se_3, *Phys. Rev. B*, **23**, pp. 2596–2606.

Vardeny, Z. and Olszakier, M. (1987). Infrared photomodulation spectroscopy of band tail states in a-Si:H and a-Si:F, *J. Non-Cryst. Solids*, **97–98**, pp. 109–112.

Vetterl, O, Finger, F., Carius, R., Hapke, P., Houben, L., Kluth, O., Lambertz, A., Mück, A., Rech, B. and Wagner, H. (2000). Intrinsic microcrystalline silicon: A new material for photovoltaics, *Sol. Energy Mat. Sol. Cells*, **62**, pp. 97–108.

Vignoli, S., Meaudre, R. and Meaudre, M. (1996). Metastable defect creation and annealing under illumination in intrinsic hydrogenated amorphous silicon deposited from helium–silane mixtures, *Philos. Mag. B*, **73**, pp. 261–276.

Watkins, G. D. and Corbett, J. W. (1964). Defects in irradiated silicon: Electron paramagnetic resonance and electron–nuclear double resonance of the Si-E center, *Phys. Rev.*, **134**, A1359–A1377.

Watkins, G. D. (1965). *A review of EPR studies in irradiated silicon, Proc. Radiation Damage in Semiconductors* (Dunod, Paris), pp. 97–113.

Watkins, G. D. (2000). *Handbook of Semiconducto Technology*, Vol. 1, eds. Jackson, K. A. and Schröter, W., Chapter 3 "Intrinsic Point Defectsin Semiconductors 1999" (Wiley-VCH, Weinheim), pp. 121–165.

Weeks, J. D., Tully, J. C. and Kimerling, L. C. (1975). Theory of recombination-enhanced defect reactions in semiconductors. *Phys. Rev. B*, **12**, pp. 3286–3292.

Wilson, D. K. and Feher, G. (1961). Electron spin resonance experiments on donors in silicon III. Investigation of excited states by the application of uniaxial stress and their importance in relaxation processes, *Phys. Rev.*, **124**, pp. 1068–1084.

Winer, K., Hirabayashi, I. and Ley, L. (1988a). Exponential conduction-band tail in P-doped a-Si:H, *Phys. Rev. Lett.*, **60**, 2697–2700.

Winer, K., Hirabayashi, I. and Ley, L. (1988b). Distribution of occupied near-surface band-gap states in a-Si:H, *Phys. Rev. B*, **38**, pp. 7680–7693.

Wraback, M. and Tauc, J. (1992). Direct measurement of the hot carrier cooling rate in a-Si:H using femtosecond 4 eV pulses, *Phys. Rev. Lett.*, **69**, pp. 3682–3685.

Yamaguchi, M. and Morigaki,K. (1993). Optical properties of amorphous-microcrystalline mixed-phase Si:H films, *J. Phys. Soc. Jpn.*, **62**, pp. 2915–2923.

Yamaguchi, M. and Morigaki, K. (1997). Photoluminescence and optically detected magnetic resonance in a-Si:H/a-Si$_3$N$_4$:H multilayers, *Phys. Rev. B*, **55**, pp. 2368–2377.

Yamaguchi, M. and Morigaki, K. (1999). Effect of hydrogen dilution on the optical properties of hydrogenated amorphous silicon prepared by plasma deposition, *Philos. Mag. B*, **79**, pp. 387–405.

Yamaguchi, M., Morigaki, K. and Nitta, S. (1989). Observation of light-induced phenomena by photoluminescence and optically detected magnetic resonance in a-Si:H, *J. Phys. Soc. Jpn.*, **58**, pp. 3828–3841.

Yamaguchi, M., Morigaki, K. and Nitta, S. (1991). Observation of light-induced phenomena by photoluminescence and optically detected magnetic resonance in a-Si$_{1-x}$N$_x$:H, *J. Phys. Soc. Jpn.*, **60**, pp. 1769–1791.

Yamasaki, S. and Isoya, J. (1993). Pulsed-ESR study of light-induced metastable defect in a-Si:H, *J. Non-Cryst. Solids*, **164–166**, pp. 169–174.

Yamasaki, S., Okushi, H., Matsuda, A. and Tanaka, K. (1990). Origin of optically induced electron-spin resonance in hydrogenated amorphous silicon, *Phys. Rev. Lett.*, **65**, pp. 756–759.

Yan, B, Schultz, N. A., Efros, A. L. and Taylor, P. C. (2000). Universal distribution of residual carriers in tetrahedrally coordinated amorphous semiconductors, *Phys. Rev. Lett.*, **84**, pp. 4180–4183.

Yokomichi, H., Hirabayashi, I. and Morigaki, K. (1987). Electron-nuclear double resonance of dangling-bond centres in a-Si:H, *Solid State Commun.*, **61**, pp. 697–701.

Yokomichi, H. and Morigaki, K. (1987). Electron-nuclear double resonance of dangling bonds centres associated with hydrogen incorporation in a-Si:H, *Solid State Commun.*, **63**, pp. 629–632.

Yokomichi, H. and Morigaki, K. (1991). The nature of structural defects and thermal equibrium effects on doped a-Si:H as elucidated by electron nuclear double resonance measurements, *J. Non-Cryst. Solids*, **137–138**, pp. 183–186.

Yokomichi, H. and Morigaki, K. (1993). Clustered dangling bonds in a-Si:H as elucidated by ENDOR measurements, *Solid State Commun.*, **85**, pp. 759–761.

Yokomichi, H. and Morigaki, K. (1996). Evidence for existence of hydrogen-related dangling bonds in hydrogenated amorphous silicon, *Philos. Mag. Lett.*, **73**, pp. 283–287.

Yonezawa, F., Sakamoto, S. and Hori, M. (1991). Ab-initio molecular dynamics simulations of amorphous silicon, *J. Non-Cryst. Solids*, **137–138**, pp. 135–140.

Yoshida, M. and Morigaki, K. (1989). Triplet exciton states in a-Si:H and its alloys as elucidated by optically detected magnetic resonance measurements, *J. Phys. Soc. Jpn.*, **58**, pp. 3371–3382.

Yoshida, M. and Taylor, P. C. (1992). Correlations of changes in ESR and PL with light soaking in a-Si:H, *MRS Proc.*, **258**, pp. 347–352.

Yoshida, M., Yamaguchi, M. and Morigaki, K. (1989). Nature of triplet excitons in a-Si:H, its alloys and a-Si:H multilayers as elucidated by optically detected magnetic resonance measurements, *J. Non-Cryst. Solids*, **114**, pp. 319–321.

Zafar, S. and Schiff, E. A. (1989). Hydrogen-mediated model for defect metastability in hydrogenated amorphous silicon, *Phys. Rev. B*, **40**, 5235–5238.

Zhang, S. B. and Branz, H. M. (2000). Nonradiative electron–hole recombination by a low-barrier pathway in hydrogenated silicon semiconductors, *Phys. Rev. Lett.*, **84**, pp. 967–970.

Zhang, Q., Nishino, T., Kumeda, M. and Shimizu, T. (1995). Light-induced annealing of photocreated dangling bonds in hydrogenated amorphous silicon, *Jpn. J. Appl. Phys.*, **34**, pp. 483–486.

Zhou, J. H., Kumeda, M. and Shimizu, T. (1995). Light-induced ESR in variously treated hydrogenated amorphous silicon, *Jpn. J. Appl. Phys.*, **34**, pp. 3982–3986.

Index